P · O · C · K · E · T · S

NATURE
FACTS

D1434279

Yellow
Mantella

Hibiscus

Thong
weed

Sugar kelp

Bladderwrack

Dulse

Carrageen

P · O · C · K · E · T · S

NATURE
FACTS

Written by
SCARLETT O'HARA

STRUCTURE
OF DNA

AMOEBA

SEAHORSES

DORLING KINDERSLEY
London • New York • Stuttgart

A DORLING KINDERSLEY BOOK

Project editor	Scarlett O'Hara
Art editor	Jacqui Burton
Senior editor	Alastair Dougall
Senior art editor	Sarah Crouch
Editorial consultant	David Burnie
Production	Josie Alabaster
Picture research	Louise Thomas

First published in Great Britain in 1997
by Dorling Kindersley Limited
9 Henrietta Street, Covent Garden, London WC2E 8PS
Reprinted 1997
Copyright © 1997 Dorling Kindersley Ltd., London

Visit us on the World Wide Web at
http://www.dk.com

A CIP catalogue record for this book is available from
the British Library.

ISBN 0 7513 5496 1

Colour reproduction by Colourscan, Singapore
Printed and bound in Italy by L.E.G.O.

CONTENTS

How to use this book

These pages show you how to use *Pockets: Nature Facts*. The book is divided into six sections. The first five provide information about plants and animals and include a section on ecology. The last section contains classification tables followed by a glossary and a comprehensive index.

Running head
Corner coding
Heading
Introduction
Diagram
Caption
Chart
ANIMALS
MOLLUSCS AND ECHINODERMS
THE SOFT BODY of a mollusc is usually covered by a hard shell. Molluscs include gastropods, such as snails; bivalves, such as oysters; and cephalopods, such as squids. Echinoderms have a five-part body and live in the sea.
INSIDE A SNAIL
TYPES OF MOLLUSC
LIFE CYCLE OF AN OYSTER

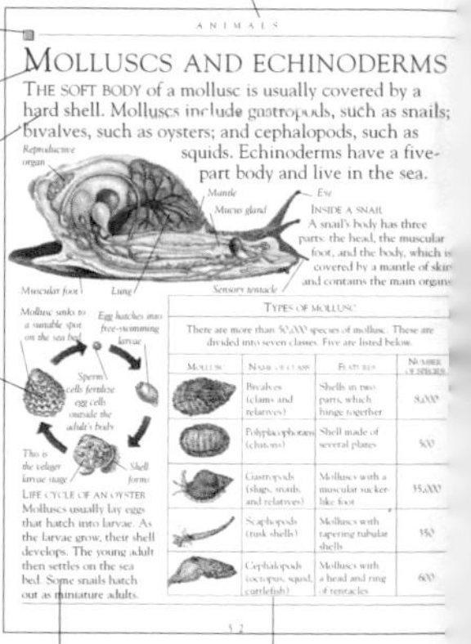

Heading and introduction

There is a heading for each spread of the book. The heading is followed by the introduction, which outlines the subject to be discussed and gives a clear idea what these pages are about.

Captions

Each image in the book, whether artwork, photograph, or diagram, is accompanied by an explanatory caption.

Corner coding

The corners of the pages in the main section of the book are colour coded. They act as a reminder of the section you are looking at.

- The beginning of life
- Microorganisms and plants
- Animals
- Animal processes
- Ecology
- Classifying living things

Diagrams

This book contains a great many diagrams. Diagrams clarify a written explanation and make information more accessible.

Charts

Charts appear on many pages in the book. They supply facts and figures. The chart above shows numbers of mollusc species.

LABELS

For extra clarity, some pictures have labels. A label identifies a picture if it is not immediately obvious what it is from the text, or it may give extra information.

RUNNING HEADS

Across the top of the pages there are running heads. The left-hand running head gives the section of the book and the right-hand the subject heading of the pages.

FEATURE BOXES

The feature boxes that appear on some pages contain detailed information and illustrations to explain a topic that is related to the main subject of the page.

Annotation

Label

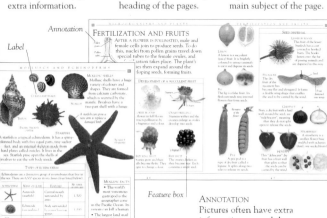

Feature box

Fact box

ANNOTATION

Pictures often have extra information around them, which picks out features. This text appears in *italics* and uses leader lines to point to details.

FACT BOXES

On many spreads there are fact boxes filled with fascinating facts and figures. The information in these is related to the main topic on the page.

INDEX AND GLOSSARY

At the back of the book is a glossary, which defines difficult words. There is also an index listing every subject in the book. By referring to the index, information on particular topics can be found quickly.

THE BEGINNING OF LIFE

WHAT IS NATURE?

Humans are the most intelligent of all animals

MOST SCIENTISTS divide nature into five kingdoms: animals, plants, fungi, and two kingdoms called Monera and Protista. Monerans and protists include single-celled forms of life, such as bacteria and algae. Animals, plants, and most fungi are multi-cellular.

HUMANS

Humans are mammals, and more specifically primates, along with apes and monkeys. Like most mammals, humans give birth to live young.

Some fungi absorb nutrients from dead matter

FUNGI AND LICHENS

Fungi and lichens were considered plants. Now, more is known about how they live, and they are thought of as a separate kingdom.

FUNGI

CHARACTERISTICS OF LIVING THINGS

- Able to use energy
- Able to take in raw materials
- Able to get rid of waste
- Able to respond to the outside world
- Able to reproduce
- Able to grow and develop

LARKSPUR

Flowering plants are very successful

PLANTS

Early plants lived in water and then adapted to conditions on land. Nearly all plants are able to use sunlight to turn simple materials into food.

GIRAFFE
WEEVIL

*Flying insects
can travel
to find food*

MONERANS AND PROTISTS
The Monera kingdom includes
the simplest forms of life, such as
bacteria. It has at least 4,000 species.
The Protista kingdom is more varied.
It includes simple forms of algae and
protozoa (animal-like organisms
such as amoebas). It comprises at
least 50,000 species.

COCCI
BACTERIA

SPIRAL
BACTERIA

AMOEBA

INSECTS
The most successful
animals on Earth are
insects. There are more
species of insect than any
other group of animals on
Earth. Insects are able to
survive in almost all
habitats except the sea.

*Mammals, like
the lion, are
warm-blooded*

ANIMALS
The animal kingdom includes
simple creatures such as sponges
and complex mammals such as lions
and humans. Animals cannot make
their own food as plants do; they must
find it and eat it. They are able to
survive in many different habitats.

LION

HOW LIFE BEGAN

THROUGHOUT HISTORY, people have wondered how life on Earth began. Some people believe it was specially created. Scientists think that the first simple life forms were the result of chemical reactions four billion years ago.

EARTH'S STEAMING SURFACE

THE BEGINNING OF THE EARTH

In the early days of the Earth, the planet was a fiery mass of molten rock. Earth's surface cracked and hot lava poured out filling the atmosphere with steam and gases. The lava hardened and the steam cooled to rain, creating steaming, muddy pools.

Surface covered with sizzling lava

Steam rises to condense and fall again as rain

MIXTURE OF GASES

Scientists have filled this flask with a mixture of gases like those that existed in the early days of the Earth. To simulate lightning, electric sparks are added. The gases combine and produce compounds that are found in all living things.

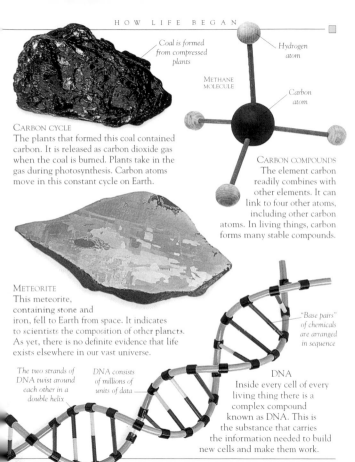

Coal is formed from compressed plants

Hydrogen atom

METHANE MOLECULE

Carbon atom

CARBON CYCLE
The plants that formed this coal contained carbon. It is released as carbon dioxide gas when the coal is burned. Plants take in the gas during photosynthesis. Carbon atoms move in this constant cycle on Earth.

CARBON COMPOUNDS
The element carbon readily combines with other elements. It can link to four other atoms, including other carbon atoms. In living things, carbon forms many stable compounds.

METEORITE
This meteorite, containing stone and iron, fell to Earth from space. It indicates to scientists the composition of other planets. As yet, there is no definite evidence that life exists elsewhere in our vast universe.

The two strands of DNA twist around each other in a double helix

DNA consists of millions of units of data

"Base pairs" of chemicals are arranged in sequence

DNA
Inside every cell of every living thing there is a complex compound known as DNA. This is the substance that carries the information needed to build new cells and make them work.

FOSSILS

THE REMAINS OF ANIMALS OR PLANTS are sometimes preserved by a process called fossilization. During this process, the hard parts of living things, such as bones, shells, and teeth, turn to rock. Fossils provide evidence (the fossil record) of how organisms have changed.

AN EARLY FOSSIL
The earliest fossils belong to the Precambrian period. They reveal groups, or colonies, of simple life forms called cyanobacteria or blue-green algae.

Discs called ossicles fit together and allow flexibility

STAURANDERASTER FOSSIL
This *Stauranderaster* is very like a modern starfish. It is covered in hard plates called ossicles that are well preserved by fossilization.

The ossicles have remained intact

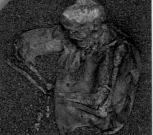

PEAT-BOG MAN
Peat bogs in northern Europe have preserved the bodies of humans for more than 2,000 years. The acidity of the peat bog prevents the complete decay of the bodies and many soft parts remain, including skin and internal organs.

Coelacanths have distinctive, lobed fins

Coelacanths can live at depths of 400 m (1,310 ft)

A LIVING "FOSSIL"

This fish, called a coelacanth, lives off the coast of southern Africa. It was thought to have been extinct for many millions of years until a fisherman caught one in 1938.

THE PROCESS OF FOSSILIZATION

Soft flesh may be eaten or decay naturally

ANIMAL BODY

Soft parts of a dead animal decay while it lies on the ground. Hard parts, such as bones, remain.

Bones covered by layers of earth

BURIED BONES

Over many years, the bones are covered by earth that is carried by wind or deposited by water.

Water deposits sand and mud

CHEMICAL CHANGES

Further deposits compress the soil and it turns to rock. The chemical make-up of the bones is altered.

Movements in the earth may expose the fossil

FOSSIL EXPOSED

Eventually, the surface of the land is worn away by weathering or erosion and the fossil is revealed.

EVOLUTION

FOSSILS SHOW THAT certain life forms have altered or "evolved" over time and that others have died out. British naturalists Charles Darwin and Alfred Russel Wallace formed the theory of evolution. It is based on "natural selection", a process that favours the best adapted members of a species.

AN EVOLUTIONARY LINK
This fossil is *Archaeopteryx* – a bird-like dinosaur with wings. It may show a link between reptiles and birds. This kind of link is important to scientists because it is evidence of how one species may have evolved from another.

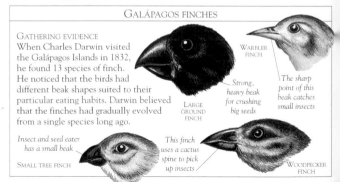

GALÁPAGOS FINCHES

GATHERING EVIDENCE
When Charles Darwin visited the Galápagos Islands in 1832, he found 13 species of finch. He noticed that the birds had different beak shapes suited to their particular eating habits. Darwin believed that the finches had gradually evolved from a single species long ago.

WARBLER FINCH

The sharp point of this beak catches small insects

Strong, heavy beak for crushing big seeds

LARGE GROUND FINCH

Insect and seed eater has a small beak

SMALL TREE FINCH

This finch uses a cactus spine to pick up insects

WOODPECKER FINCH

Hyracotherium lived 50 million years ago. It had four toes on its front feet

Mesohippus, which lived 30 million years ago, had three toes on its front feet

Merychippus existed 20 million years ago. One of its three toes formed a large hoof

Equus is the modern horse

Equus evolved about two million years ago. It has a single toe or hoof on each foot

EVOLUTION OF THE HORSE
The earliest fossils of a horse-like animal, called *Hyracotherium*, show a small mammal, about the size of a dog, that ate leaves. Over time, descendants became larger with longer legs and a different diet. Modern horses are larger still and eat grass.

NATURAL SELECTION
Darwin believed evolution favoured individuals that were best suited to their environment. Less successful individuals of the same species would naturally die out. In 19th-century England, dark peppered moths became more common than pale peppered moths. Pollution meant that dark moths were better camouflaged from birds on the blackened trees.

PEPPERED MOTHS ON LIGHT BARK

ARTIFICIAL SELECTION
The size, colour, and shape of animals such as horses, dogs, and cats can be modified artificially through selective breeding. Breeders choose individuals with the desired qualities and reject the rest.

Sphynx is bred to be hairless

SPHYNX CAT

Human evolution

Apes are humans' closest relatives. Apes and humans together are known as hominoids. Humans and their direct ancestors are called hominids. The earliest hominid fossils are from about 3.5 million years ago. One such fossil, found in Ethiopia, is called "Lucy". She had a small brain but walked upright. The most recent hominids belong to the group *Homo* and appeared about two million years ago. These include *Homo habilis*, who used tools.

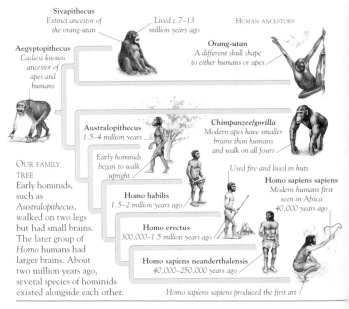

Sivapithecus
Extinct ancestor of the orang-utan

Lived c.7–13 million years ago

HUMAN ANCESTORS

Aegyptopithecus
Earliest known ancestor of apes and humans

Orang-utan
A different skull shape to either humans or apes

Australopithecus
1.5–4 million years

Early hominids began to walk upright

Chimpanzee/gorilla
Modern apes have smaller brains than humans and walk on all fours

OUR FAMILY TREE
Early hominids, such as *Australopithecus*, walked on two legs but had small brains. The later group of *Homo* humans had larger brains. About two million years ago, several species of hominids existed alongside each other.

Homo habilis
1.5–2 million years ago

Used fire and lived in huts

Homo sapiens sapiens
Modern humans first seen in Africa 40,000 years ago

Homo erectus
300,000–1.5 million years ago

Homo sapiens neanderthalensis
40,000–250,000 years ago

Homo sapiens sapiens produced the first art

SKELETON SHAPES
Gorillas walk on their feet and hands. The human skeleton is adapted for upright walking. It has a forward-pointing big toe, whereas the gorilla's toe is angled for grasping. Human hip bones are smaller to make striding easier.

Toe positioned for holding onto trees

Toe points forwards on human foot

GORILLA SKELETON

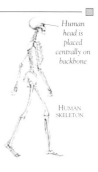

Human head is placed centrally on backbone

HUMAN SKELETON

FIRST HUMAN FOOTPRINTS
Mary Leakey, a British fossil expert, unearthed these fossilized footprints in Africa in 1977. They were left by three hominids in volcanic ash nearly four million years ago.

Apes evolved 30 million years ago

Hominids appeared only around 5 million years ago

Dinosaurs lived 200 million years ago

Earth was formed 4.6 billion years ago

Reptiles appear

If you imagine Earth's history in one hour, then humans appear just before the hour strikes

Land plants emerge

Bacteria appeared 3.8 billion years ago

Plant life in the oceans began 1.5 billion years ago

ARRIVAL TIMES OF HUMANS	
Humans originated in Africa, but spread to many other regions much more recently.	
REGION	APPROXIMATE ARRIVAL TIME
Australia	At least 40,000 years ago
North America	At least 12,000 years ago
Madagascar	2,000 years ago
New Zealand	1,000 years ago
Antarctica	150 years ago

EVOLUTIONARY CLOCK
Scientists believe simple life forms first appeared on Earth 3.8 billion years ago. It was not until about five million years ago that the first hominids appeared.

GEOLOGICAL TIMESCALE

EARLY LIFE FORMS appeared on Earth about 3,800 million years ago. More complex plants and animals evolved in the last 600 million years. Animals such as the dinosaurs have died out and other animals have taken their place. Earth itself also changes. Its rocky crust shifts and its climate alters.

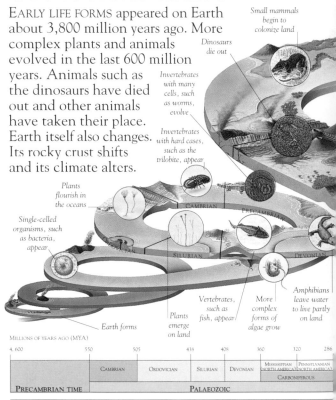

Small mammals begin to colonize land

Dinosaurs die out

Invertebrates with many cells, such as worms, evolve

Invertebrates with hard cases, such as the trilobite, appear

Plants flourish in the oceans

Single-celled organisms, such as bacteria, appear

Vertebrates, such as fish, appear

More complex forms of algae grow

Amphibians leave water to live partly on land

Plants emerge on land

Earth forms

MILLIONS OF YEARS AGO (MYA)

4,600	550	505	438	408	360	320	286
	CAMBRIAN	ORDOVICIAN	SILURIAN	DEVONIAN	MISSISSIPPIAN (NORTH AMERICA)	PENNSYLVANIAN (NORTH AMERICA)	
					CARBONIFEROUS		
PRECAMBRIAN TIME		PALAEOZOIC					

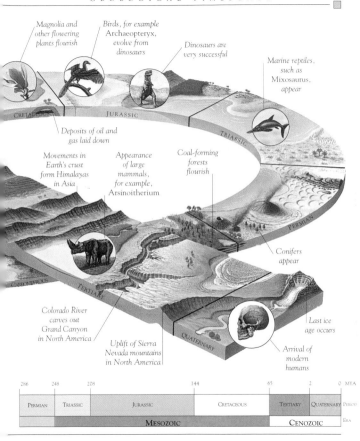

Magnolia and other flowering plants flourish

Birds, for example Archaeopteryx, evolve from dinosaurs

Dinosaurs are very successful

Marine reptiles, such as Mixosaurus, appear

CRETACEOUS

JURASSIC

TRIASSIC

Deposits of oil and gas laid down

Movements in Earth's crust form Himalayas in Asia

Appearance of large mammals, for example, Arsinoitherium

Coal-forming forests flourish

PERMIAN

Conifers appear

CARBONIFEROUS TERTIARY

Colorado River carves out Grand Canyon in North America

Uplift of Sierra Nevada mountains in North America

QUATERNARY

Last ice age occurs

Arrival of modern humans

286	248	208		144		65	2	0 MYA
PERMIAN	TRIASSIC	JURASSIC		CRETACEOUS		TERTIARY	QUATERNARY	PERIODS
		MESOZOIC				CENOZOIC		ERA

MICROORGANISMS AND PLANTS

MICROSCOPIC LIFE

THERE ARE MANY kinds of life that are so tiny they are invisible except through a microscope. These include billions of bacteria and viruses, some of which live in the human body. Another group of small organisms, called protists, also consist of just a single cell.

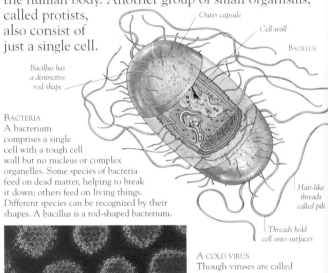

Outer capsule

Cell wall

BACILLUS

Bacillus has a distinctive rod shape

BACTERIA
A bacterium comprises a single cell with a tough cell wall but no nucleus or complex organelles. Some species of bacteria feed on dead matter, helping to break it down; others feed on living things. Different species can be recognized by their shapes. A bacillus is a rod-shaped bacterium.

Hair-like threads called pili

Threads hold cell onto surfaces

A COLD VIRUS
Though viruses are called microorganisms, they are not strictly living things because they cannot reproduce without the help of living cells. In animals, viruses can cause colds and flu or diseases such as AIDS.

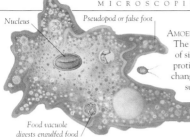

Nucleus

Pseudopod or false foot

Food vacuole digests engulfed food

AMOEBA

The jelly-like amoeba belongs to a group of single-celled organisms known as protists. Amoebas move by constantly changing shape. They take in food by surrounding and engulfing it. Food is stored and digested in a structure called a food vacuole.

CELLS

ANIMAL CELL

The flexible skin around an animal cell allows some chemicals into the cell but not others. The cell is controlled by the nucleus. In the fluid around the nucleus, called cytoplasm, lie tiny organelles that perform specific tasks.

Plasma membrane

Cytoplasm

Nucleus – the cell's command centre

Energy-producing organelle

Chloroplasts trap the energy in sunlight

Plasma membrane

Cell wall

Vacuole used to store cell sap

Nucleus

PLANT CELL

A plant cell differs from an animal cell in two important ways. It has an additional rigid cell wall as well as a plasma membrane around it and it contains green-coloured organelles called chloroplasts.

Slug forms into fruiting body

Fruiting body releases spores

Spores germinate

Amoeba grows from spore

Amoebas come together into slug

Slug migrates

LIFE OF A SLIME MOULD

A slime mould reproduces in an unusual way. First, slime mould amoebas gather together to form a mass called a slug. The slug moves towards the light. Eventually, it forms a tall, fruiting body that releases spores. The spores produce new amoebas.

FUNGI AND LICHENS

FUNGI ARE NEITHER plants nor animals but members of a separate kingdom that includes mushrooms, moulds, and yeasts. Many fungi exist as tiny threads that form large fruiting bodies. Lichens are living partnerships between fungi and algae.

Fruiting body releases spores

Spores form hyphae

Network called mycelium

Mycelium spreads

Fruiting body begins to form

LIFE CYCLE OF A FUNGUS

For most of its life a fungus is a network of threads, called hyphae, in the fungus's food source. Fungi form fruiting bodies to release their spores.

Brightly coloured, poisonous cap

FLY AGARIC TOADSTOOL (POISONOUS)

Spores form in gills

Stalk is made up of many threads meshed together

BUDDING YEAST

Microscopic fungi called yeast exist as individual cells. During their reproduction process, known as budding, new cells begin as buds that grow and eventually separate from the parent cell.

TOADSTOOL

The life cycle of most fungi includes a fruiting body such as a toadstool or mushroom. The fruiting body provides a way for the fungi to reproduce. Spores (seed-like specks) are made in gills under the cap and released when conditions are right.

PUFFBALL
A puffball fungus starts out as a solid head of cells. Gradually, the cells dry out and form a papery bag. When drops of rain or an animal brush the bag, a cloud of spores is released into the air. The spores are very light and can be carried long distances.

The fruiting body of the puffball puffs out spores

FUNGI FACTS
• The fungus *Penicillium notatum* produces the well-known antibiotic drug, penicillin.

• A giant puffball can release seven million million spores in its life.

• The seeds of many orchids will not germinate without the presence of a fungus.

TYPES OF LICHEN

A lichen is not a plant, but a partnership between a fungus and an alga that exists as one organism. Lichens grow on rocks and tree trunks. The fungus dissolves substances that the alga uses; and the alga supplies the fungus with food made by photosynthesis. Lichens grow in five forms, of which three – leafy (foliose), crusty and flat (crustose), and mixed (squamulose) – are shown here.

Leafy type of lichen

Tree bark

Lichens will eventually dissolve minerals from the rock

CLADONIA FLOERKEANA

Spore-producing body

HYPOGYMNIA PHYSODES

Flat and crusty lichens

CALOPLACA HEPPIANA

Mixture of shrubby and leafy lichen

ALGAE AND SIMPLE PLANTS

PLANTS EVOLVED FROM ALGAE, which consist of either one cell or many cells. Algae do not have true roots or leaves, and they also lack flowers. Liverworts and mosses were some of the earliest plants on Earth. They grow in damp, shady places. Ferns and horsetails were among the first plants to develop water-carrying systems. They are known as vascular plants.

EXAMPLES OF ALGAE

New colonies growing

Adult colony

VOLVOX
Volvox is a freshwater alga. It consists of many cells and is often found in ponds.

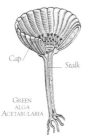

Cap

Stalk

GREEN ALGA ACETABULARIA

ACETABULARIA
This is a single-celled, cup-shaped alga found in shallow sea water.

SEAWEEDS
A seaweed is a plant-like marine alga made of many cells. It usually anchors itself to the sea bed near the shore, and its leaf-like fronds trap the energy in sunlight.

Kelp is a brown seaweed

Bladderwrack has air bladders to make it float on the surface

Sugar kelp has fronds – long, leaf-like flaps

Carrageen

Dulse is a red seaweed

Liverworts grow close to the ground

LIVERWORT

Liverworts usually grow in damp areas where there is shade. Like mosses, liverworts are bryophytes, a group of simple plants that have no true stems, leaves, or roots. Liverworts are flat, ribbon-like plants. Their life cycle, like that of ferns, has two stages.

Moss grows on rotting wood

MOSSY LOG

Other simple plants, mosses, live mostly in damp places. Mosses do not have true roots but only thread-like rhizoids. Mosses are also non-vascular, meaning that they lack a system for carrying water.

LIFE CYCLE OF A FERN

The life cycle of a fern has two quite different stages. During the first stage, a leafy plant (the sporophyte) creates spores. These produce the small, heart-shaped second stage (the gametophyte). This makes male and female cells that fuse, forming a new sporophyte.

Side branches in circles from the stem

Sporophyte or green plant produces spores

Released spores grow into gametophytes

Gametophyte

Male and female cells (gametes) produced

Fertilization creates new sporophyte

FERN FROND

HORSETAIL

Common in damp places, horsetails are an ancient, brush-like plant. They spread by spores and have creeping underground stems.

31

CONIFERS

Needles grow in rosettes

PINES, CEDARS, AND CYPRESSES are examples of conifers. Conifers form part of a group of plants called gymnosperms. These were the first land plants to reproduce with seeds rather than spores. Conifers usually grow their seeds in hard, woody cones. Most conifers, including firs and pines, are evergreen, which means that they keep their leaves throughout the year.

BLUE ATLAS CEDAR

Narrow needles to reduce water loss

Triangular, hard, spiky leaves

Needles have a thick outer layer and a wax coating

AROLLA PINE

SCALES AND NEEDLES
Conifer leaves tend to be narrow with a waxy coating to reduce moisture loss. The long, thin leaves of pines, firs, and cedars are called needles, the short, flat leaves of cypresses are called scales. Most conifer leaves are dark green and feel leathery. Conifers, particularly pines, are well adapted to survival in dry or freezing conditions.

Bunches of needles in groups of five

MONKEY PUZZLE

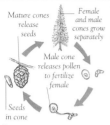

LIFE CYCLE OF A CONIFER
A conifer grows male and female cones. Pollen from male cones fertilizes female cones that later form seeds.

Mature cones release seeds

Female and male cones grow separately

Male cone releases pollen to fertilize female

Seeds in cone

TALLEST AND HEAVIEST TREES

TALLEST TREE
Coast redwoods are the world's tallest trees, reaching 111 m (364 ft).

HEAVIEST TREE
Another redwood, the giant sequoia, is the world's heaviest tree.

Mature female cone sits upright on tree

CEDAR CONES

Cones break up over a few months

CEDAR CONE
The seed scales inside the mature cedar cone peel away and the cone disintegrates while still on the tree.

SCOTS PINE CONE
The mature female cone of the Scots pine opens up and the seeds flutter away into the air.

AMBER
Some conifers produce resin to protect themselves from insects, which get trapped in the sticky sap. Prehistoric insects have been preserved in fossilized resin, called amber.

FLOWERING PLANTS

THE MOST SUCCESSFUL PLANTS on Earth are the flowering plants. They can survive in many kinds of habitat, from mountainsides to deserts. There are more than 250,000 species of flowering plants and nearly all produce food by a process called photosynthesis. Most of these plants have green leaves and live in soil.

Leaves reduced to spines

Flowers form distinctive, five-petal shape

CACTUS FAMILY
Cactus family (Cactaceae) plants live in dry places. They have no leaves and store water in their stems.

PEA FAMILY
The pea family (Leguminosae) includes the food plants, beans and peas. The flowers have five petals and produce a fruit known as a legume. The legume holds one or more seeds.

FLOWERING PLANT FACTS

• The Tallipot palm *Corypha umbraculifera* takes about 100 years to flower and then dies.

• Potatoes, peppers, and tomatoes are members of the nightshade family, which also includes many poisonous plants.

GRASS FAMILY
The most wide-spread flowering plants are grasses (Gramineae). Many are grown for their seeds (grain), such as wheat, rice, maize, and oats.

BEECH FAMILY
The beech family (Fagaceae) includes trees such as sweet chestnuts, oaks, and beeches. Their fruits are nuts and their flowers are frequently catkins. In the northern hemisphere beech trees are used as a source of timber.

Flowers have five petals

Thorns grow on stems

Many pollen-producing stamens

ROSE FAMILY
In the rose family, (Rosaceae) many plants, such as roses, apples, cherries, and strawberries, are cultivated. Some rose family species are trees or shrubs. The flowers usually have four or five petals and many stamens at the centre.

Adapted flowers attract animals

ORCHID FAMILY
The second-largest family of flowering plants is the orchid family (Orchidaceae). Their specialized flowers attract animal pollinators. Orchids rely on a special relationship with fungi to grow.

LARGEST FAMILIES OF FLOWERING PLANTS		
SCIENTIFIC NAME	COMMON NAME	NUMBER OF SPECIES
Compositae	Daisy family	25,000
Orchidaceae	Orchid family	18,000
Leguminosae	Pea family	17,000
Gramineae	Grass family	9,000
Rubiaceae	Coffee family	7,000
Euphorbiaceae	Spurge family	5,000
Cyperaceae	Rush family	4,000

Plants and leaves

A plant's stem supports its buds, leaves, and flowers. Inside the stem are cells that carry water, minerals, and food to different parts of the plant. Roots anchor the plant and absorb water and minerals from the soil. Leaves use the Sun's energy to make food.

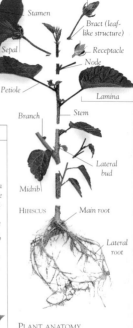

Stigma
Petal
Flower bud
Stamen
Bract (leaf-like structure)
Sepal
Receptacle
Node
Petiole
Lamina
Branch
Stem
Lateral bud
Midrib
HIBISCUS
Main root
Lateral root

LEAVES

LEAF SHAPES
A simple leaf has a single blade (lamina), but a compound leaf is divided into separate leaflets. In a bipinnate compound leaf, the leaflets are divided.

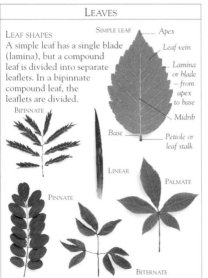

SIMPLE LEAF
Apex
Leaf vein
Lamina or blade – from apex to base
Midrib
Base
Petiole or leaf stalk

BIPINNATE

LINEAR

PALMATE

PINNATE

BITERNATE

PLANT ANATOMY
The visible part of the plant is the shoot. It consists of a stem supporting buds (undeveloped shoots), leaves, and flowers. The shoot grows towards the light. Roots fix the plant to the ground and collect water.

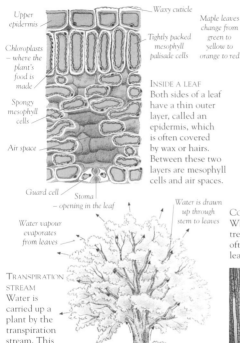

Upper epidermis

Waxy cuticle

Chloroplasts – where the plant's food is made

Tightly packed mesophyll palisade cells

Spongy mesophyll cells

Air space

Guard cell

Stoma – opening in the leaf

INSIDE A LEAF
Both sides of a leaf have a thin outer layer, called an epidermis, which is often covered by wax or hairs. Between these two layers are mesophyll cells and air spaces.

Maple leaves change from green to yellow to orange to red

COLOURFUL LEAVES
When leaves of deciduous trees die, chemical changes often occur, turning the leaves bright colours.

Water is drawn up through stem to leaves

Water vapour evaporates from leaves

TRANSPIRATION STREAM
Water is carried up a plant by the transpiration stream. This constant flow of water replaces moisture lost through openings in the leaves called stomata. Water is drawn into the plant's roots and pulled up through the stem.

Water passes into roots from soil

XYLEM AND PHLOEM
In the stem, xylem cells form tubes that carry water and minerals from the roots to the leaves. Phloem cells take food to all parts.

PHOTOSYNTHESIS

PLANTS MAKE THEIR FOOD by a process known as photosynthesis. It requires sunlight, water, and carbon dioxide. Photosynthesis is carried out mostly in the leaves, where the pigment that gives leaves their green colour, called chlorophyll, is stored. Oxygen is released during this process.

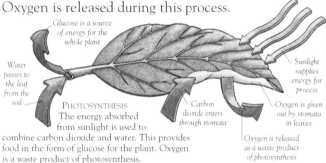

Glucose is a source of energy for the whole plant

Water passes to the leaf from the soil

Sunlight supplies energy for process

Oxygen is given out by stomata in leaves

Carbon dioxide enters through stomata

Oxygen is released as a waste product of photosynthesis

PHOTOSYNTHESIS
The energy absorbed from sunlight is used to combine carbon dioxide and water. This provides food in the form of glucose for the plant. Oxygen is a waste product of photosynthesis.

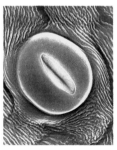

PHOTOSYNTHESIS FACTS

• If photosynthesis stopped, almost all life on Earth would cease.

• Each year, plants make about 100 billion tonnes of glucose by harnessing the light of the Sun.

• Earth's atmosphere would contain no oxygen without photosynthesis.

STOMATA
On the underside of leaves are microscopic pores called stomata. These openings allow the gases carbon dioxide and oxygen to pass into and out of the leaf. Guard cells are situated on either side of each pore and control its opening and closing.

INSIDE A CHLOROPLAST

Photosynthesis takes place in structures called chloroplasts. These are inside plant cells, mostly in the leaves. They contain chlorophyll, a green pigment that absorbs energy from sunlight.

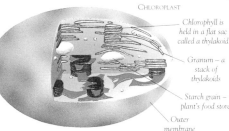

CHLOROPLAST

Chlorophyll is held in a flat sac called a thylakoid

Granum – a stack of thylakoids

Starch grain – plant's food store

Outer membrane

Daytime: plant gives out more oxygen than it uses

Twilight: plant releases equal amounts of gases

Night-time: plant releases more carbon dioxide

CHLOROPHYLL

When sunlight reaches a plant, red and blue light is absorbed and green light is reflected, making the plant appear green. The pigment chlorophyll, in chloroplasts, reflects the green colour.

PHOTOSYNTHESIS AND RESPIRATION

During the day, plants make food by photosynthesis. They take in carbon dioxide and release oxygen. At night, photosynthesis stops and plants respire. They break down food, taking in oxygen and releasing carbon dioxide.

VENUS FLY TRAP

Some plants living in acid, boggy soils make food by photosynthesis but get some nutrients from animals. The Venus fly trap has a spring-trap mechanism that allows it to catch insects.

Enzymes inside leaf digest insects

Trap sealed with interlocking spines

VENUS FLY TRAP

Lobes of trap formed from modified leaves

Insects nudge trigger hairs on the edge of the trap

PLANT GROWTH

WHEN A PLANT BEGINS TO GROW, its shoot reaches up so its leaves can absorb sunlight, and its roots grow down to absorb water and nutrients from the soil. Once a plant is mature, it reproduces by flowering or in other ways. Many plants store food for the next growing season.

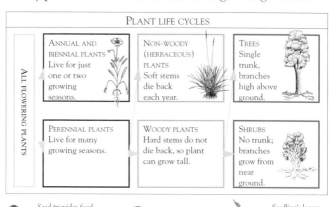

PLANT LIFE CYCLES

ALL FLOWERING PLANTS

ANNUAL AND BIENNIAL PLANTS
Live for just one or two growing seasons.

NON-WOODY (HERBACEOUS) PLANTS
Soft stems die back each year.

TREES
Single trunk, branches high above ground.

PERENNIAL PLANTS
Live for many growing seasons.

WOODY PLANTS
Hard stems do not die back, so plant can grow tall.

SHRUBS
No trunk; branches grow from near ground.

Seed provides food for young plant

Roots

Leaves begin to photosynthesize

Shoot

Seedling's leaves now make food to fuel growth

1 GERMINATION
Once the seed coat (testa) has split and the roots have begun to grow down, the first shoot reaches up towards the light.

2 GROWING
Growth takes place at the tips of the seedling's roots and shoot. The shoot reaches upwards.

3 SEEDLING
Leaves provide the plant with fuel by photosynthesis. Roots take in nutrients and water from the soil.

REPRODUCING WITHOUT SEEDS

Many plants, including this strawberry, can reproduce without making seeds. To do this, they grow parts that can take root and grow. This is asexual reproduction, because only one parent is involved.

Parent plant reproduces in two ways – it also makes seeds

New plant begins to grow in different place

Parent plant sends out runners over the ground

STORAGE SYSTEMS

SWEET POTATO

Plants such as the sweet potato store food from one growing season to the next. The food is held in a swollen underground stem called a tuber.

GRAPE HYACINTH

The bulb produced by plants such as the grape hyacinth is another type of storage system. Food is stored in layers of fleshy scales packed into the bulb.

A bulb is a type of short underground stem

HOW TREES GROW

TREE RINGS

Wood is a strong material that supports trees and shrubs. It consists of layers of xylem cells toughened with a substance called lignin. Each growing season a new ring of xylem is added.

Young tree has smooth bark

Cracked bark of mature tree

BARK

The layer on the outside of a woody plant's stem is known as bark.
The outer layer of bark is dead, but under this there are living phloem cells. Tiny pores in the bark, called lenticels, allow gases to pass through.

FLOWERS

FLOWERS ENABLE A PLANT to be pollinated and then to form seeds. Flowers consist of male and female organs surrounded by petals that may be coloured and scented to attract visiting animals. The male parts of a flower (stamens) make pollen; the female parts (carpels) make cells that form seeds.

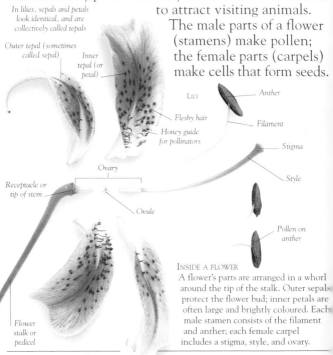

In lilies, sepals and petals look identical, and are collectively called tepals

Outer tepal (sometimes called sepal)

Inner tepal (or petal)

LILY

Fleshy hair

Honey guide for pollinators

Anther

Filament

Ovary

Receptacle or tip of stem

Ovule

Stigma

Style

Pollen on anther

Flower stalk or pedicel

INSIDE A FLOWER

A flower's parts are arranged in a whorl around the tip of the stalk. Outer sepals protect the flower bud; inner petals are often large and brightly coloured. Each male stamen consists of the filament and anther; each female carpel includes a stigma, style, and ovary.

GIANT FLOWER

The giant rafflesia (*Rafflesia arnoldii*) is the world's largest flower. It can grow up to 105 cm (3.5 ft) across and weigh up to 7 kg (15.4 lb). It is at risk from the destruction of its habitat in the rainforests of Sumatra.

CAPITULUM

UMBEL

SPIKE

RACEME

INFLORESCENCE

Inflorescences are groups of flowers arranged on a single stem. Sunflowers and their relatives form a capitulum, which contains many flowers but looks like one flower. A spike consists of stalkless flowers on a straight stem; a raceme has flowers with stalks. An umbel is umbrella-shaped.

ENDANGERED FLOWERING PLANTS

Of the 300,000 or so species identified, about six per cent are endangered. Plants are at risk from over-collection or destruction of their habitat. This is a selection of key endangered species.

CONTINENT	SPECIES	THREATENED BY
North America	Saguaro cactus *Carnegia gigantea*	Over-collection
South America	Chilean wine palm *Juba chilensis*	Land clearance
Africa	African violet *Saintpaulia ionantha*	Forest clearance
Asia	Orchid *Paphiopedilum rothschildianum*	Over-collection
Europe	Madonna lily *Lilium candidum*	Over-collection
Australia	Lobster claw *Clianthus puniceus*	Browsing from livestock

POLLINATION

POLLINATION ENSURES that a plant can develop seeds and so reproduce. The process can be carried out by animals, usually insects, or by the wind or water. Pollen is transferred from the male part of one flower to the female part of another. Both flowers must be of the same species.

SUNFLOWER UNDER ULTRAVIOLET LIGHT

Outer petals look pale

ANIMAL POLLINATION

When an insect, such as a bee, visits a flower, pollen from the flower's anthers is brushed onto it. The bee collects nectar and pollen to take to the hive, and may visit other flowers of the same species, transferring pollen grains between flowers as it feeds.

ATTRACTING INSECTS

Seen under ultraviolet light, the centre of this sunflower looks dark and the petals look pale. Honeybees see this contrast, and are attracted to the nectar-rich, dark centre of the flower.

BUMBLEBEE ON A DOG ROSE

Pollen is stored in sacs on the bee's legs

Bee collects nectar from inside the flower

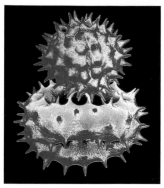

WIND POLLINATION
The flowers of wind-pollinated plants are often small and dull. When the flowers are mature, their dry, dusty pollen is scattered by the wind to other flowers. Many grasses, in particular, are wind-pollinated.

SWEET CHESTNUT
(MALE)

Filament

Anther

Hanging
flowers are
called catkins

POLLEN GRAINS
The pollen grains above are enlarged many times. Unmagnified, they look like particles of yellow dust. Pollen is produced by the anthers of a flower. It contains the male sex cells.

Sticky pollen grains
attach themselves
to feathers

HUMMINGBIRD
FROM THE
AMAZON
RAINFOREST

HUMMINGBIRD
Some flowers are pollinated by birds or bats. A hummingbird drinks nectar from inside a flower and picks up pollen grains on its feathers. It then carries this pollen to other flowers.

POLLINATION FACTS
• The smallest pollen grains are only 0.02 mm (0.0008 in) across.

• Ribbonweed, a water plant, has pollen that floats in "boats" to female flowers.

• A species of orchid in Madagascar is pollinated by a moth with a 30 cm (1 ft) long tongue.

FERTILIZATION AND FRUITS

AFTER A FLOWER IS POLLINATED, male and female cells join to produce seeds. To do this, nuclei from pollen grains travel down special tubes to the female ovules, and fertilization takes place. The plant's ovaries then expand around the developing seeds, forming fruits.

Pollen grain

Pollen tube

Ovule

Ovum (female sex cell)

FERTILIZATION

Fertilization occurs after a pollen grain containing male cells reaches a flower's stigma. A tube grows down from the grain to the ovule. Male nuclei travel along the tube to join a female cell.

SEED FACTS

• The coco-de-mer palm *Laodicea maldivica* has the largest seeds. Each fruit weighs up to 20 kg (44 lb) and has one seed.

• Orchids have the smallest seeds. A billion can weigh 1 g (0.035 oz).

DEVELOPMENT OF A SUCCULENT FRUIT

FLOWER BLOOMS
A flower in full bloom attracts pollinators by its fragrance and colour.

Stamen

Ovary

OVARY SWELLING
Stamens wither and the ovaries enlarge as ovules develop into seeds.

Ovary

MATURING FRUIT
Ovaries grow and their walls become fleshy. They begin to change colour.

Drupelet

RIPE FRUIT
The ovaries darken as they become ripe. Each contains a single seed.

LEMON

A lemon is a succulent (juicy) fruit. It is brightly coloured to attract animals to eat it and disperse its seeds.

FIG

The fig is a false fruit. Its flesh surrounds tiny internal flowers that form seeds.

SEED DISPERSAL

LESSER BURDOCK

This fruit of the lesser burdock has a coat covered in hooked burrs. The hooks fasten onto the fur of passing animals and are dispersed in this way.

SYCAMORE

Seed

The dry fruit of the sycamore has become flat and elongated. It forms a double wing shape that enables the seed to be carried by the wind.

Fruit flattened into wings

Remains of the flower

CHESTNUT

Nuts, a dry fruit with a hard wall around the seed, are "indehiscent", meaning that they do not split open to release the seeds.

STRAWBERRY

A strawberry is a swollen flower base studded with achenes (hard, one-seeded fruits).

PEA

A pea pod is a type of dry fruit called a legume. It splits along two sides to release its seeds.

HONESTY

This "dehiscent" dry fruit has a fruit wall that splits so that the seeds can be carried by the wind.

ANIMALS

SIMPLE INVERTEBRATES

AN ANIMAL WITHOUT a backbone is known as an invertebrate. Invertebrates make up over nine-tenths of all animal species. Some are simple, microscopic creatures; others larger and more complex.

Digestive cavity in the middle of the bell

Plankton trapped in mucus

Adult jellyfish

Fleshy arms collect food

The ephyrae break free

JELLYFISH
A jellyfish moves by contracting its bell-shaped body. It has trailing tentacles that pull food into the digestive cavity, which is in the middle of the jellyfish.

Polyp splits into eight-armed buds called ephyrae

Small polyps form

LIFE CYCLE OF A JELLYFISH
Adult jellyfish release fertilized larvae that rest on the sea bed. The larvae grow into small polyps that divide into buds and swim away as tiny adults.

MUSHROOM CORAL
FUNGIA

CORAL
A coral is a polyp with a cylindrical shape. Most corals live together in colonies. Corals may build hard cases of calcium carbonate. When they die these may form coral reefs.

TYPES OF SIMPLE INVERTEBRATE			
There are more than a million known species of invertebrate divided into 30 phyla (types). These are two of the simplest.			
INVERTEBRATE	NAME OF PHYLUM	FEATURES	NUMBER OF SPECIES
	Poriferans (sponges)	Simple animals that filter food from water	5,000
	Cnidarians (sea anemones, corals, jellyfish)	Simple animals with stinging threads	10,000

TAPEWORM AND ITS HEAD

Hooks and suckers on the head

EARTHWORM

An earthworm has a segmented body, which it moves using two sets of muscles. The muscles change the shape of the segments. Earthworms feed on the organic matter in soil. An earthworm is both male and female (hermaphroditic).

Gizzard (part of the stomach)

Earthworms have several "hearts"

Body is formed from identical segments

Mouth

INSIDE AN EARTHWORM

The worm releases eggs in packets

TAPEWORM

A tapeworm is a type of parasitic flatworm that lives inside the intestines of animals such as pigs, cats, or even humans. The worm's head has hooks and suckers that grasp the intestines. Food is absorbed from the host animal's gut.

TYPES OF INVERTEBRATE (WORMS)			
Worms are often divided up into repeated segments. Most are free-living, but some are parasitic.			
INVERTEBRATE	NAME OF PHYLUM	FEATURES	NUMBER OF SPECIES
	Annelids (worms, leeches)	Worms with a segmented body	12,000
	Platyhelminthes (flatworms, flukes, tapeworms)	Worms with a simple flattened body	10,000
	Nematodes (roundworms)	Worms with an unsegmented cylindrical body	15,000

MOLLUSCS AND ECHINODERMS

THE SOFT BODY of a mollusc is usually covered by a hard shell. Molluscs include gastropods, such as snails; bivalves, such as oysters; and cephalopods, such as squids. Echinoderms have a five-part body and live in the sea.

Reproductive organ

Mantle

Mucus gland

Eye

INSIDE A SNAIL
A snail's body has three parts: the head, the muscular foot, and the body, which is covered by a mantle of skin and contains the main organs.

Muscular foot

Lung

Sensory tentacle

Mollusc sinks to a suitable spot on the sea bed

Egg hatches into free-swimming larvae

Sperm cells fertilize egg cells outside the adult's body

The veliger larvae stage

Shell forms

LIFE CYCLE OF AN OYSTER
Molluscs usually lay eggs that hatch into larvae. As a larva grows, its shell develops. The young adult then settles on the sea bed. Some snails hatch out as miniature adults.

TYPES OF MOLLUSC			
There are more than 50,000 species of mollusc. These are divided into seven classes. Five are listed below.			
MOLLUSC	NAME OF CLASS	FEATURES	NUMBER OF SPECIES
	Bivalves (clams and relatives)	Shells in two parts, which hinge together	8,000
	Polyplacophorans (chitons)	Shell made of several plates	500
	Gastropods (slugs, snails, and relatives)	Molluscs with a muscular sucker-like foot	35,000
	Scaphopods (tusk shells)	Molluscs with tapering tubular shells	350
	Cephalopods (octopus, squid, cuttlefish)	Molluscs with a head and ring of tentacles	600

CUBAN LAND SNAILS

TRITON

PACIFIC THORNY OYSTER

SCALLOP

MOLLUSC SHELLS

Mollusc shells have a huge variety of colours and shapes. They are formed from calcium carbonate, which is secreted by the mantle. Bivalves have a two-part shell with a hinge.

A starfish can grow a new arm to replace a damaged limb

SCARLET STARFISH

STARFISH

A starfish is a typical echinoderm. It has a spiny-skinned body with five equal parts, tiny sucker feet, and an internal skeleton made from hard plates called ossicles. It lives in the sea. Starfish prise open the shells of bivalves to eat the soft body inside.

TYPES OF ECHINODERM

Echinoderms are a distinctive group of invertebrates that live in the sea. There are 6,500 species in six classes (four listed below).

ECHINODERM	NAME OF CLASS	FEATURES	NUMBER OF SPECIES
	Asteroids (starfish)	Central mouth surrounded by arms	1,500
	Echinoids (sea urchins)	Body surrounded by a case bearing spines	1,000
	Crinoids (feather stars)	Mouth surrounded by feathery arms	600
	Holothuroidea (sea cucumbers)	Worm-like body with feeding tentacles	1,100

MOLLUSC FACTS

• The world's most venomous gastropod is the geographer cone in the Pacific Ocean. Its venom can kill a human.

• The largest land snail is the giant African land snail *Achatina achatina*. It can grow up to 39 cm (15.4 in) from head to tail.

ARTHROPODS

ARACHNIDS, CRUSTACEANS, and insects are part of the arthropod group of invertebrates. Insects are by far the largest of these three groups. All arthropods have a jointed body with a tough body case. The case is shed as the animal grows.

The egg is laid in a silk sac to protect it

Spiderlings resemble the adult spider

Spiderling moults

LIFE CYCLE OF A SPIDER
Arachnids such as spiders lay eggs that hatch into tiny versions of adults. They moult several times before they are mature.

IMPERIAL SCORPION

Poison gland
Sting
Heart
Intestine
Cephalothorax
Pedipalps – a pair of pincers for feeding
Abdomen
Spiracle – air hole

INSIDE AN ARACHNID
The body of an arachnid is divided into a front and middle part (cephalothorax) and a rear part (abdomen). Arachnids have four pairs of walking legs.

TYPES OF ARACHNID					
The class Arachnida includes spiders, mites, and scorpions. It contains 73,000 species, which are grouped into ten orders. Six orders are listed below.					
ARACHNID	NAME OF ORDER	NUMBER OF SPECIES	ARACHNID	NAME OF ORDER	NUMBER OF SPECIES
	Scorpiones (scorpions)	2,000		Uropygi (whip scorpions)	60
	Solifugae (camel spiders)	900		Opiliones (harvestmen)	4,500
	Acari (mites and ticks)	30,000		Araneae (spiders)	40,000

INSIDE A CRUSTACEAN
A typical crustacean has a hard
body case with a head, thorax,
and abdomen. It has compound
eyes, two pairs of antennae,
and many pairs of jointed legs.

Antenna

COMMON
LOBSTER

*Cephalothorax (joined
head and thorax)*

Heart

Abdomen

GOLIATH
BEETLE

*Strong legs
used for fighting*

*Shell or
carapace*

Elytron

Intestine

Swimmerets

A BEETLE
The Goliath beetle
is the world's heaviest.
It weighs up to 100 g (3.5 oz). Beetles are
very well armoured. Their forewings are hardened,
curved plates called elytra. The elytra protect
the fragile hind wings that are used for flying.

TYPES OF CRUSTACEAN			
There are more than 55,000 species of crustaceans divided into eight classes. These include the four classes below.			
CRUSTACEAN	NAME OF CLASS	FEATURES	NUMBER OF SPECIES
	Branchiopods (fairy shrimps, water fleas)	Small animals of freshwater and salty lakes	1,000
	Cirripedia (barnacles)	Immobile animals with a box-like case	1,220
	Copepods (cyclopoids and relatives)	Small animals often found in plankton	13,000
	Malacostracans (shrimps, crabs, lobsters)	Many-legged animals, often with pincers	30,000

*Mature
adult*

*Egg is fertilized
outside body*

*First
larval
stage*

*Post-
larval
stage*

*Second
larval stage*

LIFE CYCLE OF A SHRIMP
Crustaceans usually lay
their eggs in water. Once
hatched, the egg begins
its first larval stage. After
two more larval stages,
there is a final post-larval
stage before adulthood.

INSECTS

INSECTS MAKE UP over four-fifths of all the animal species on Earth. About 800,000 species are known, and many thousands more are discovered each year. Insects live in almost every habitat on land, from rainforest to desert. Many live in fresh water, but hardly any live in the sea. Most insects can fly, and many change shape as they mature.

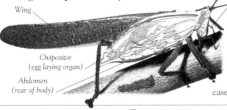

Antenna

KATYDID (FEMALE)

Brain

Mandibles (mouthparts)

Wing

Ovipositor (egg laying organ)

Abdomen (rear of body)

INSIDE
AN INSECT
An insect's body is split into three parts: the head, the thorax, and the abdomen. It has six jointed legs, a hard body case, and it usually has wings.

TYPES OF INSECT					
Insects are part of the phylum Arthropoda. There are 800,000 known species of insect in the class Insecta. They are grouped into 32 orders, six of which are shown below.					
INSECT	NAME OF ORDER	NUMBER OF SPECIES	INSECT	NAME OF ORDER	NUMBER OF SPECIES
	Odonata (dragonflies)	5,000		Hemiptera (bugs)	82,000
	Orthoptera (grasshoppers and crickets)	20,000		Coleoptera (beetles)	350,000
	Lepidoptera (butterflies and moths)	170,000		Hymenoptera (ants, bees, and wasps)	110,000

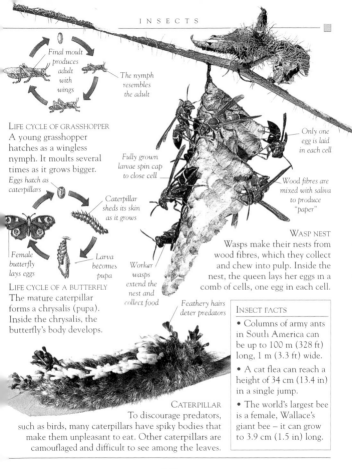

LIFE CYCLE OF GRASSHOPPER
A young grasshopper
hatches as a wingless
nymph. It moults several
times as it grows bigger.

*Final moult
produces
adult
with
wings*

*The nymph
resembles
the adult*

*Eggs hatch as
caterpillars*

*Caterpillar
sheds its skin
as it grows*

*Female
butterfly
lays eggs*

*Larva
becomes
pupa*

LIFE CYCLE OF A BUTTERFLY
The mature caterpillar
forms a chrysalis (pupa).
Inside the chrysalis, the
butterfly's body develops.

*Only one
egg is laid
in each cell*

*Fully grown
larvae spin cap
to close cell*

*Wood fibres are
mixed with saliva
to produce
"paper"*

*Worker
wasps
extend the
nest and
collect food*

WASP NEST
Wasps make their nests from
wood fibres, which they collect
and chew into pulp. Inside the
nest, the queen lays her eggs in a
comb of cells, one egg in each cell.

*Feathery hairs
deter predators*

CATERPILLAR
To discourage predators,
such as birds, many caterpillars have spiky bodies that
make them unpleasant to eat. Other caterpillars are
camouflaged and difficult to see among the leaves.

INSECT FACTS

• Columns of army ants
in South America can
be up to 100 m (328 ft)
long, 1 m (3.3 ft) wide.

• A cat flea can reach a
height of 34 cm (13.4 in)
in a single jump.

• The world's largest bee
is a female, Wallace's
giant bee – it can grow
to 3.9 cm (1.5 in) long.

FISH

WITH MORE THAN 20,000 species, fish are more numerous than all other vertebrates (animals with backbones). Fish are very well suited to life in the water. They have streamlined bodies, and most have slippery scales and a special organ that helps them float. There are three distinct types of fish; bony fish are the most widespread.

INSIDE A FISH
Bony fish have a skeleton made of bone. They swim using their tail and are covered by slimy scales. Fish have gills to obtain oxygen and a gas-filled swim bladder to keep them buoyant.

CRUCIAN CARP (FEMALE)

Dorsal fin
Backbone
Swim bladder
Spinal cord
Gill arch
Gill slits
Caudal fin
Mouth
Anal fin
Pelvic fin
Pectoral fin
Heart

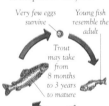

Very few eggs survive
Young fish resemble the adult
Trout may take from 8 months to 3 years to mature

LIFE CYCLE OF A TROUT
Some fish give birth to fully formed young but most release eggs, often thousands at a time. The young fish are called fry.

TYPES OF FISH			
There are more than 20,000 species of fish. These are divided into three groups. The largest group is made up of bony fish.			
FISH	NAME OF CLASS	FEATURES	NUMBER OF SPECIES
	Agnatha (jawless fish)	Fish with sucker-like mouths	75
	Chondrichthyes (cartilaginous fish)	Fish with skeletons of cartilage	800
	Osteichthyes (bony fish)	Fish with skeletons of bone	22,000

CAMOUFLAGE
Flatfish such as plaice or sole gradually alter their skin colour and pattern to blend in with their sea-bed background.

SEAHORSES
The seahorse is an unusual fish. It swims upright and holds onto weeds with its grasping tail. Females lay their eggs in a pouch on the male's abdomen.

LOACH

UPSIDE-DOWN CATFISH

Mountain lakes and streams
Fish such as loach and salmon can live at altitudes of up to 4,900 m (16,000 ft).

Lakes and rivers
In freshwater lakes and rivers live carp, characins, and catfish.

FISH HABITATS
Fish can live almost anywhere there is water. They have adapted to life in the deepest oceans as well as along the shore and even in underwater caves.

Shoreline
Along the shoreline are fish, such as the mudskipper, that spend some time out of water.

MUDSKIPPER

Coastal waters
Coral reefs in tropical coastal waters are home to many brightly coloured fish.

MANDARIN FISH

PACIFIC MANTA RAY

Open ocean
In the open ocean, fish such as rays and skates can grow to a large size. Sharks hunt in this environment.

Caves
Some cave-dwelling fish have no eyes because they spend their lives in darkness.

BLIND CAVE CHARACIN

Deep ocean
The large mouths and expanding stomachs of deep-sea fish trap more of the food available.

GULPER EEL

Middle-ocean depths
With little sunlight reaching these depths and the water cooler, few fish live here.

OARFISH

AMPHIBIANS

AMPHIBIANS WERE THE FIRST GROUP of animals to move from water to live on land. Most amphibians spend the early part of their lives in water. Later, they grow legs, lose their gills, and can live both on land and in water.

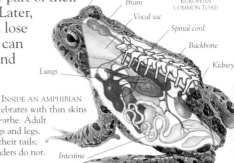

Eye
Brain
EUROPEAN COMMON TOAD
Vocal sac
Spinal cord
Backbone
Kidney
Lungs
Intestine

INSIDE AN AMPHIBIAN
Amphibians are vertebrates with thin skins through which they breathe. Adult amphibians have lungs and legs. Frogs and toads lose their tails; newts and salamanders do not.

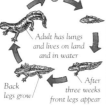

Adult has lungs and lives on land and in water

Back legs grow

After three weeks front legs appear

LIFE CYCLE OF A NEWT
Larvae, called tadpoles, hatch out in water. They breathe through gills. After about eight weeks, the tadpoles develop legs and their gills disappear.

TYPES OF AMPHIBIAN

There are about 3,000 species of amphibian that make up the class Amphibia. They are divided into three orders.

AMPHIBIANS	NAME OF ORDER	NUMBER OF SPECIES
	Apoda (caecilians)	170
	Anura (frogs and toads)	3,700
	Urodela (newts and salamanders)	350

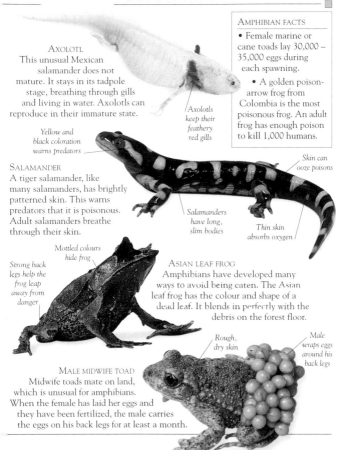

AXOLOTL
This unusual Mexican salamander does not mature. It stays in its tadpole stage, breathing through gills and living in water. Axolotls can reproduce in their immature state.

Axolotls keep their feathery red gills

Yellow and black coloration warns predators

SALAMANDER
A tiger salamander, like many salamanders, has brightly patterned skin. This warns predators that it is poisonous. Adult salamanders breathe through their skin.

Salamanders have long, slim bodies

Skin can ooze poisons

Thin skin absorbs oxygen

Mottled colours hide frog

Strong back legs help the frog leap away from danger

ASIAN LEAF FROG
Amphibians have developed many ways to avoid being eaten. The Asian leaf frog has the colour and shape of a dead leaf. It blends in perfectly with the debris on the forest floor.

Rough, dry skin

Male wraps eggs around his back legs

MALE MIDWIFE TOAD
Midwife toads mate on land, which is unusual for amphibians. When the female has laid her eggs and they have been fertilized, the male carries the eggs on his back legs for at least a month.

REPTILES

MILLIONS OF YEARS AGO, reptiles dominated Earth. Some of these reptiles – including dinosaurs – died out, but other kinds are still alive today. Unlike amphibians, reptiles may live entirely on dry land, but they are also cold-blooded, and so need warmth to become active. Most reptiles lay eggs.

INSIDE A REPTILE
Reptiles are vertebrates with scaly skins. Except for snakes and some lizards, most reptiles have four legs. A reptile's skin is tough and waterproof, and helps it to retain water in hot, dry conditions.

Nostril
Brain
Heart
Stomach
Bladder
Kidney
Backbone

EYED-LIZARD
(FEMALE)

TYPES OF REPTILE

There are about 6,000 species of reptile in the class Reptilia. These are divided into four orders.

REPTILE	NAME OF ORDER	NUMBER OF SPECIES
	Squamata (lizards and snakes)	5,700
	Crocodilia (crocodilians)	23
	Rhynchocephalia (tuataras)	2
	Chelonia (turtles, tortoises, and terrapins)	200

Egg sealed with membrane to prevent drying out

Young resembles adult

Mature adult continues to grow

LIFE CYCLE OF A GECKO
Many reptiles lay eggs that are sealed with a strong membrane. The young hatch out as miniature versions of their parents.

CROCODILES
The ancient Crocodilia order includes crocodiles, alligators, and gavials, which have narrower jaws. Crocodiles spend a lot of time in water and can shut their nostrils when they dive. Despite a vicious appearance (sharp teeth, strong jaws, and a powerful tail) they are careful parents.

Crocodiles carry their young in their mouth

FEMALE NILE CROCODILE

BOA CONSTRICTOR
Pythons, boa constrictors, and anacondas are snakes that feed mainly on mammals that they kill by constriction. The snake coils around its prey, and as the animal struggles, it squeezes tighter until the animal suffocates. Then the snake swallows it.

The snake opens its special hinged jaws to swallow its victim

KOMODO DRAGON

Wrinkly, loose fitting skin

LARGEST LIZARD
The world's largest lizard is a monitor lizard called the Komodo dragon. It can grow up to 3 m (10 ft) long and weigh up to 166 kg (365 lb). There are fewer than 5,000 Komodo dragons left. They live on a few islands in Indonesia, including Komodo Island.

BIRDS

BIRDS ARE THE LARGEST ANIMALS capable of powered flight. Their streamlined bodies are covered with feathers, and many bones are hollow to save weight. Birds live in different habitats, and reproduce by laying hard-shelled eggs. Some species have lost the ability to fly.

Brain

Backbone

Beak

Crop – to store food temporarily

Stomach

SPECKLED PIGEON

Liver

Foot with claws

Rectum

Tail feathers

INSIDE A BIRD
Birds are warm-blooded vertebrates. Their feathers give them a lightweight, warm covering. They have beaks (or bills) and lay eggs that have hard shells.

TYPES OF BIRD

About 9,000 species of bird have been discovered. These divide into 28 orders. Perching birds (Passeriformes) make up the largest order (5,414 species). Six other orders are listed here.

BIRD	NAME OF ORDER	NUMBER OF SPECIES	BIRD	NAME OF ORDER	NUMBER OF SPECIES
	Rheiformes (rheas)	2		Anseriformes (waterfowl)	150
	Piciformes (woodpeckers, toucans, barbets)	381		Falconiformes (birds of prey)	290
	Psittaciformes (parrots, lories, cockatoos)	342		Strigiformes (owls)	174

Eggs have hard shells

Chick uses special tooth to escape egg

Bird has adult plumage after six weeks and can then fly

Chick has soft down feathers

LIFE CYCLE OF A MOORHEN
A moorhen's eggs are incubated by both parents. After 21 days, the chicks hatch, using a special tooth to chip their way out. The young can swim and feed a few hours after hatching.

TYPES OF FEATHER

FEATHER FUNCTIONS
Feathers are strong and flexible. They are made of strands of a protein called keratin. The strands are often hooked together to form a flat surface. A bird has several kinds of feathers, suited to particular purposes.

WING FEATHER

DOWN FEATHER

Soft, fluffy down keeps bird warm

Tail feathers are long and stiff for flight

These cover bird's body and keep it streamlined

Barbs hook together to keep surface flat

CONTOUR FEATHERS

TAIL FEATHER

TYPES OF BEAK

BEAKS FOR A PURPOSE
A bird's beak, or bill, is bone with a covering of horn. The shape of the beak is suited to the bird's feeding habits. Birds of prey, such as kestrels, have sharp, hooked beaks for tearing flesh and macaws have beaks with a pointed end to pierce fruit.

GREEN-WINGED MACAW

KESTREL

GREATER FLAMINGO

AVOCET

Lower beak pumps water through upper sieve in beak

Upturned beak swings back and forth in water

MAMMALS

MAMMALS ARE A VERY DIVERSE GROUP of animals.
They range from tiny shrews to huge rhinos and from
whales that spend their lives in the oceans to bats that
spend time in the air. Mammals can live in nearly all
habitats – jungles, rivers,
and deserts. Humans
belong to this group
of animals.

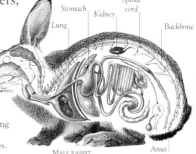

Spinal cord

Stomach   Kidney

Lung   Backbone

Brain

Nasal cavity

Anus

MALE RABBIT

INSIDE A MAMMAL
All mammals are warm-
blooded. Most mammals have
hair of some sort on their bodies.
Female mammals suckle their young
with their own supply of milk. A
mammal has well-developed senses.

TYPES OF MAMMAL					
There are about 4,000 species in the class Mammalia. They are divided into 21 orders. The largest order is the Rodentia, which includes 2,021 species of rodent. Six orders are listed here.					
MAMMAL	NAME OF ORDER	NUMBER OF SPECIES	MAMMAL	NAME OF ORDER	NUMBER OF SPECIES
	Monotremes (egg-laying mammals)	3		Artiodactyla (even-toed, hoofed mammals)	220
	Marsupials (mammals that grow in a pouch)	272		Carnivora (meat eaters)	237
	Insectivora (insect eaters)	428		Cetacea (whales and dolphins)	78

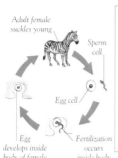

Adult female suckles young

Sperm cell

Egg cell

Fertilization occurs inside body

Egg develops inside body of female

MARSUPIALS

Wallabies, kangaroos, and koalas are marsupials. Their young are born in a very immature state after a short gestation time. They crawl into the mother's pouch and spend time suckling there while they develop.

LIFE CYCLE OF A ZEBRA

Sperm cells from the male must join an egg cell in the female for fertilization to take place. This occurs inside the female mammal. Most mammals develop in the womb of their mother and all are fed on her milk.

INDIAN FRUIT BAT

FLYING MAMMALS

Bats are the only mammals capable of powered flight. A bat flies by flapping a thin membrane of skin stretched between each of its front and back legs. The largest bat may have a wingspan of 2 m (6.5 ft).

Mother squirrel suckles young for first few weeks

SQUIRREL WITH YOUNG

A mammal whose baby develops inside the mother's womb (uterus) is known as a placental mammal. Inside the womb, the babies are protected while they develop and are fed with oxygen and food via the placenta.

Young are well-developed at birth and grow rapidly

Baby squirrels can soon find their own food

PRIMATES

THIS GROUP OF MAMMALS includes monkeys, apes, and humans. The primates form the most intelligent group of animals. With forward-facing eyes for judging distances, grasping fingers, and long limbs, primates are ideally suited to life in the trees. Some primates also have prehensile tails that grip onto branches.

PATAS MONKEY

Closely set nostrils of Old World monkey

OLD WORLD MONKEY

Monkeys from the Old World, which includes Europe, Africa, and Asia, have long noses and closely set nostrils. The Patas monkey lives in groups of up to 20 females and one male on the plains, where it eats grass and seeds.

SQUIRREL MONKEY

NEW WORLD MONKEY

New World monkeys live in scrub and forest in Central and South America. Most of these monkeys have flat faces and nostrils set wide apart, and many also have prehensile tails that can wrap around branches like an extra limb. Howler and squirrel monkeys are both New World monkeys.

LORIS

Animals such as lemurs, lorises, and bushbabies are known as prosimians, or primitive primates. They are almost all nocturnal (active at night) and they have excellent eyesight. Most eat leaves, insects, and birds.

Large eyes provide good night vision

Grasping hands cling to branches

GORILLA

The fearsome-looking gorilla is actually a vegetarian. Gorillas walk on their feet and front knuckles and spend most of their time on the ground. Some gorillas live in lowland rainforests and others in the mountains.

ORANG-UTAN
Like gorillas, chimpanzees, and gibbons, orang-utans are not monkeys but apes. These shy and solitary primates live in the rainforests of Borneo and Sumatra, and spend all their life in the trees.

AVERAGE PRIMATE WEIGHTS

The 233 species of primate vary in size from the tiny pygmy marmoset to the mighty gorilla and between males and females.

SPECIES	AVERAGE WEIGHT (FEMALE)	AVERAGE WEIGHT (MALE)
Gorilla	105 kg (231 lb)	205 kg (452 lb)
Human	52 kg (115 lb)	75 kg (165 lb)
Proboscis monkey	9 kg (20 lb)	19 kg (42 lb)
Patas monkey	5.5 kg (12 lb)	10 kg (22 lb)
Pygmy marmoset	120 g (4.2 oz)	140 g (5 oz)

ANIMAL PROCESSES

NUTRITION AND DIGESTION

ANIMALS LIVE BY TAKING IN FOOD and breaking it down into simple substances. It fuels their muscles and body processes and provides the raw materials for growth. Herbivores are animals that feed only on plants. Carnivores eat meat, while omnivores eat a variety of foods.

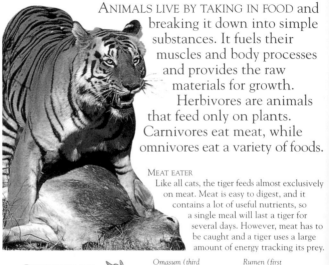

MEAT EATER

Like all cats, the tiger feeds almost exclusively on meat. Meat is easy to digest, and it contains a lot of useful nutrients, so a single meal will last a tiger for several days. However, meat has to be caught and a tiger uses a large amount of energy tracking its prey.

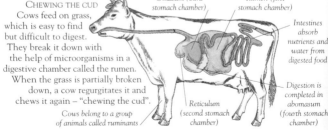

CHEWING THE CUD

Cows feed on grass, which is easy to find but difficult to digest. They break it down with the help of microorganisms in a digestive chamber called the rumen. When the grass is partially broken down, a cow regurgitates it and chews it again – "chewing the cud".

Cows belong to a group of animals called ruminants

Omasum (third stomach chamber)

Rumen (first stomach chamber)

Intestines absorb nutrients and water from digested food

Digestion is completed in abomasum (fourth stomach chamber)

Reticulum (second stomach chamber)

FLUID FEEDERS

Many adult insects survive entirely on liquid food. They include mosquitoes that feed on blood, and aphids and cicadas that feed on plant sap. Butterflies feed mainly on nectar from flowers, using a long drinking tube that coils up when not used.

Sugary nectar is a good source of energy

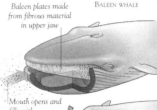

BALEEN WHALE

Baleen plates made from fibrous material in upper jaw

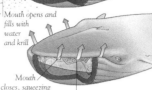

Mouth opens and fills with water and krill

Mouth closes, squeezing water through baleen plates

Krill trapped in comb-like baleen plates

FILTER FEEDING

Small animals called krill are abundant in the sea but are difficult to collect in any quantity. Whales filter them from the water using rows of baleen plates.

FEEDING FACTS

• Insects eat unusual foods. Certain moths in Southeast Asia feed on the tears of wild cattle.

• Parasitic tapeworms lack a digestive system. They absorb digested food from their hosts.

• Cold-blooded animals, such as spiders, may live for weeks without food.

EGG-EATING SNAKE

Some snakes specialize in a diet of eggs. An African egg-eating snake works its jaws around its food, and swallows it. As the egg travels down the snake's throat, bony spines break it open. It spits out the broken shell.

TEETH AND JAWS

ANIMALS USE THEIR TEETH to grip their food and to cut or chew it so that it is easier to digest. Teeth, brought together by powerful jaw muscles, are tough enough to withstand pressure without breaking. Human teeth stop growing when they are fully formed, but some animals' teeth – such as those of rodents – grow all their life.

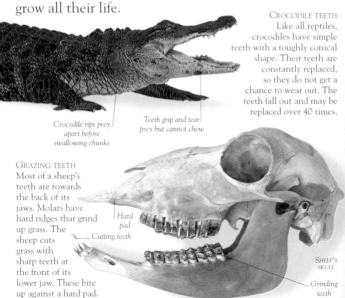

CROCODILE TEETH
Like all reptiles, crocodiles have simple teeth with a roughly conical shape. Their teeth are constantly replaced, so they do not get a chance to wear out. The teeth fall out and may be replaced over 40 times.

Crocodile rips prey apart before swallowing chunks

Teeth grip and tear prey but cannot chew

GRAZING TEETH
Most of a sheep's teeth are towards the back of its jaws. Molars have hard ridges that grind up grass. The sheep cuts grass with sharp teeth at the front of its lower jaw. These bite up against a hard pad.

Hard pad

Cutting teeth

Wear

SHEEP'S SKULL

Grinding teeth

THE DENTAL TOOLKIT

A VARIETY OF TEETH
Unlike other animals, mammals have several kinds of teeth that work together. In a dog, incisors and canines grab hold of food. Carnassials slice through it, and molars and premolars chew the food.

DOG'S TEETH

MOLAR   CARNASSIAL   PREMOLAR   INCISOR

CANINE

BITING AND CUTTING TEETH
A dog usually has a total of 42 teeth (10 more than an adult human). For dogs, biting is more important than chewing, and most of their teeth have sharp points or edges

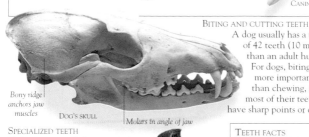

Bony ridge anchors jaw muscles

DOG'S SKULL   Molars in angle of jaw

SPECIALIZED TEETH
Tusks are unusually long teeth that protrude beyond a mammal's jaws. Elephant tusks are specialized incisors, teeth that other mammals use for biting. Elephants use them mainly to dig up food, to strip bark from trees, and for fighting.

Upper part of tusk is solid

Lower part of tusk has a soft pulpy centre

ELEPHANT'S TUSKS

TEETH FACTS

• Shark's teeth constantly move to the edge of its jaw. Each one falls out after about two weeks to be replaced by one behind it.

• An adult elephant chews with four huge teeth. The teeth are replaced slowly. This stops when 24 teeth have been replaced.

• Turtles and tortoises are completely toothless.

BREATHING

ALL ANIMALS NEED TO TAKE IN OXYGEN, and at the same time they have to get rid of a waste gas called carbon dioxide. Very small animals do this through the surface of their bodies, but most larger animals do it with the help of special organs, such as gills or lungs.

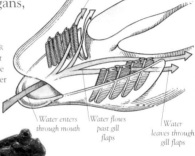

Water flows out

BREATHING IN WATER
A fish's gills are arches that support stacks of thin flaps, which are supplied with blood. When water flows past the flaps, oxygen travels into the blood and carbon dioxide flows into the water.

Water enters through mouth

Water flows past gill flaps

Water leaves through gill flaps

Mudskippers are careful to stay damp

FISH OUT OF WATER
Mudskippers are small fish that live in mangrove swamps. They can survive underwater as well as in air. In the air, they breathe by taking gulps of water to keep some water in their gills. They probably absorb some oxygen through their gills and some through the lining of their mouths.

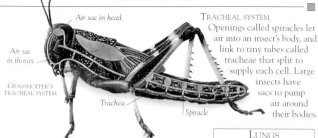

GRASSHOPPER'S
TRACHEAL SYSTEM

Air sac in head

Air sac
in thorax

Trachea

Spiracle

TRACHEAL SYSTEM

Openings called spiracles let air into an insect's body, and link to tiny tubes called tracheae that split to supply each cell. Large insects have sacs to pump air around their bodies.

LUNGS

Air in

Intercostal
muscles

INHALATION

Intercostal muscles pull ribs up while diaphragm moves down; ribcage expands and air flows in.

BIRD'S ONE-WAY LUNGS

Birds need a lot of oxygen to fly. They are very efficient at extracting it from the air they breathe. Special air sacs allow air to move straight through their lungs, which ensures they get as much oxygen as possible from the air.

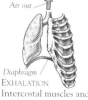

Trachea
(windpipe)

Air sacs

Lungs

Air out

EXHALATION

Intercostal muscles and diaphragm relax, lungs and ribcage deflate. Air is forced out of lungs.

Diaphragm

BIRD'S RESPIRATORY SYSTEM

COMING UP FOR AIR

Like all mammals, whales have lungs and they breathe air. They have a "blowhole" on the top of their head rather than nostrils. Some whales can hold their breath for over an hour when they dive.

Blowhole

BLOOD AND CIRCULATION

BLOOD DELIVERS SUBSTANCES that cells need, and takes away their waste. In simple animals, such as molluscs and insects, it flows mainly through open spaces inside the body, but in vertebrates, including mammals, it flows through a system of tubes called blood vessels. Blood is pumped around by the heart.

EUROPEAN LOBSTER

BLUE-BLOODED CRUSTACEAN
In many animals, blood is coloured red because it contains a red substance called haemoglobin that carries oxygen. In lobsters and other crustaceans, the oxygen-carrier is blue.

Lobster's blood contains a blue pigment – haemocyanin

MOVING HEAT
As well as moving dissolved chemicals, blood also moves heat. When a lizard basks in the sunshine, the blood beneath its skin warms up. The blood carries this heat to its internal organs.

Blood absorbs heat from skin

Body begins to warm up, enabling lizard to become active

OCELLATED LIZARD BASKING IN SUN

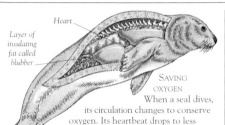

Heart

Layer of insulating fat called blubber

CIRCULATION FACTS

• Earthworms have five pairs of hearts, close to the front of their bodies.

• During hibernation, many animals' blood temperature drops close to 0° C (32° F).

• In a lifetime, a human heart pumps enough blood to fill 100 full-sized swimming pools.

SAVING OXYGEN

When a seal dives, its circulation changes to conserve oxygen. Its heartbeat drops to less than a tenth of its normal rate, and many of its blood vessels close, so that only vital organs – such as its brain – receive any blood.

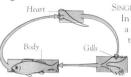

Heart

Body

Gills

SINGLE CIRCULATION

In fish, blood flows in a single circuit. From the heart, it flows through the gills, collects oxygen, and flows on to the body.

Lungs

CIRCULATION THROUGH LUNGS

Heart pumps blood around two circuits

Blood returns to other side of heart

CIRCULATION AROUND BODY

ANIMAL HEART RATES

Animals with small bodies have much faster heartbeats than larger ones.

ANIMAL	BEATS PER MINUTE
Grey whale	9
Harbour seal (diving)	10
Elephant	25
Human	70
Harbour seal (at surface)	140
Sparrow	500
Shrew	600
Hummingbird (hovering)	1,200

DOUBLE CIRCULATION

In humans and other mammals, blood makes two circuits, and the heart is a double pump. In the first circuit, it flows from the heart to the lungs and back. In the second, it flows around the rest of the body.

Body

SKELETONS

ANIMALS SUCH AS JELLYFISH have completely soft bodies that work well in water, but create problems on land. Other animals have a solid framework, or skeleton. This gives them their shape, and provides something for their muscles to pull against. Skeletons can be inside the body or surround it from the outside.

Armoured thorax

Tubular legs

AN OUTSIDE SKELETON
An insect's body is covered by an exoskeleton made of separate plates that hinge together at flexible joints. The advantage of this skeleton is that it is very tough, and helps to stop insects from drying out. To grow, an insect sheds its exoskeleton from time to time and grows a larger one in its place.

Hard forewing (elytron)

All organs are inside the skeleton

Separate plates on abdomen

JEWEL BEETLE'S EXOSKELETON

NAUTILUS SHELL

AN OUTER SHELL
A shell is a hard case that protects a soft-bodied animal. Unlike an insect's exoskeleton, it can grow, so the animal inside never has to shed it. This nautilus shell has gas-filled compartments to help the mollusc control its depth in water.

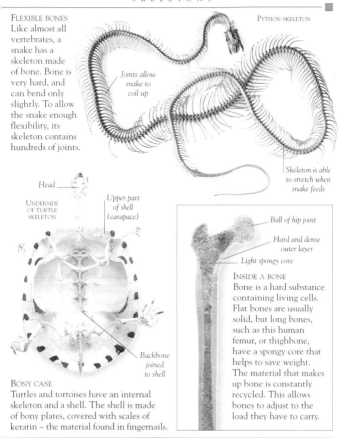

FLEXIBLE BONES
Like almost all vertebrates, a snake has a skeleton made of bone. Bone is very hard, and can bend only slightly. To allow the snake enough flexibility, its skeleton contains hundreds of joints.

PYTHON SKELETON

Joints allow snake to coil up

Skeleton is able to stretch when snake feeds

Head

UNDERSIDE OF TURTLE SKELETON

Upper part of shell (carapace)

Backbone joined to shell

BONY CASE
Turtles and tortoises have an internal skeleton and a shell. The shell is made of bony plates, covered with scales of keratin – the material found in fingernails.

Ball of hip joint

Hard and dense outer layer

Light spongy core

INSIDE A BONE
Bone is a hard substance containing living cells. Flat bones are usually solid, but long bones, such as this human femur, or thighbone, have a spongy core that helps to save weight. The material that makes up bone is constantly recycled. This allows bones to adjust to the load they have to carry.

More skeletons

The skeletons of vertebrates, or animals with backbones, all follow the same underlying plan, but over millions of years they have become modified for many different ways of life. For some animals – particularly birds – lightness is important; for others, such as elephants, the emphasis is on strength; and for others, such as fish, flexibility is vital.

Paper-thin skull

Highly flexible neck

CROW SKELETON

Ribs anchor muscles used in swimming

COD SKELETON

Backbone

Keel anchors wing muscles

SUPPORTED IN WATER

A fish's body is buoyed up by water, so its skeleton does not have to be as strong as that of an animal that lives on land. Fish have a large number of vertebrae, and many pairs of thin and flexible ribs.

Slender leg bones

LIGHT BONES

Compared to other land animals, birds have lightweight skeletons with few bones. Many of the bones are honeycombed with air spaces (reducing bones' weight), and some of these spaces connect to the air sacs that a bird uses to breathe.

FOUR-LEGGED SKELETON

A salamander's skeleton shows the typical, four-legged plan that evolved long ago, when vertebrates first took up life on land. The salamander's legs are small and weak and splay outwards.

Large skull

JAPANESE SALAMANDER SKELETON

Long backbone

Small legs

Salamander often rests with its body on ground

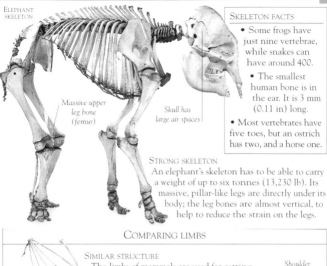

ELEPHANT
SKELETON

*Massive upper
leg bone
(femur)*

*Skull has
large air spaces*

SKELETON FACTS

• Some frogs have just nine vertebrae, while snakes can have around 400.

• The smallest human bone is in the ear. It is 3 mm (0.11 in) long.

• Most vertebrates have five toes, but an ostrich has two, and a horse one.

STRONG SKELETON

An elephant's skeleton has to be able to carry a weight of up to six tonnes (13,230 lb). Its massive, pillar-like legs are directly under its body; the leg bones are almost vertical, to help to reduce the strain on the legs.

COMPARING LIMBS

SIMILAR STRUCTURE

The limbs of mammals are used for getting around. They are all based on the same bone arrangement. Our arms are long, slender, and flexible. A bat's "arms" have evolved into wings, with long fingers that spread in flight. A porpoise's forelimbs are short and strong.

*Long
fingers
support
wing*

BAT'S WING

*Shoulder
blade
(scapula)*

*Long
forefinger*

PORPOISE'S
FLIPPER

*Long fingers
and flexible
thumb*

*Fingers are hidden
inside flippers*

*Short
arm bones*

*Mobile
shoulder*

HUMAN ARM

MUSCLES AND MOVEMENT

THE ABILITY TO MOVE is a sign of life in the animal world. Animals move by using muscles. These contain special cells that can contract or relax, making parts of the body change their position or shape. Muscles need energy to work, and animals get this from food.

Frog is propelled forwards

Powerful muscles extend back legs

Body is streamlined in flight

JUMPING FROG
Frogs, fleas, and kangaroos move by jumping. A frog has strong muscles in its back legs that contract to produce enough leverage to push the animal through the air.

MOTION FACTS

• A flea can jump 100 times its own height using energy stored in pads in its leg joints. A cat flea can leap 34 cm (13.4 in).

• A sea gooseberry moves through water by beating clusters of tiny hairs called cilia.

HUMAN MUSCLES

A pair of muscles raises the forearm

Triceps relaxes

RELAXED
Muscles pull but they do not push. They are often found in pairs where they work against each other. When the arm is at rest, the muscles are relaxed.

Biceps contracts

CONTRACTED
To raise the forearm, the biceps muscle contracts and the triceps relaxes. To lower the arm again, the triceps contracts and the biceps relaxes.

FASTEST MAMMALS

The cheetah is the fastest animal on land over short distances.
Over long distances, the pronghorn antelope is faster.

COMMON NAME	SCIENTIFIC NAME	MAXIMUM SPEED KM/H	MPH
Cheetah	Acinonyx jubatus	105	65
Pronghorn antelope	Antilocapra americana	86	53
Mongolian gazelle	Procapra gutturosa	80	50
Springbok	Antidorcas marsupialis	80	50
Grant's gazelle	Gazella granti	76	47
Thomson's gazelle	Gazella thomsoni	76	47
European hare	Lepus capensis	72	45

Limpet is held fast by its muscular foot

STUCK FAST
A mollusc such as a limpet
has special muscles that stay
contracted for hours and
use up very little energy.
The muscles are used to
close the mollusc's shell,
or to clamp it to a rock.

ROUGH LIMPET

SIDEWINDER MOVEMENT
Most snakes curve from
side to side as they push
against the ground. This
is known as serpentine
movement. The
sidewinder is different.
It throws itself forwards,
leaving a distinctive trail.

FOUR-LEGGED MOVEMENT
Animals with four legs
move in a coordinated way.
To walk, a cheetah moves its
front right leg and its rear left
leg forwards and then the opposite
pair. However, when it runs, its
front legs move together and then
its back legs move together.

A cheetah can reach 96 km/h (60 mph) in three seconds

CHEETAH RUNNING

Animal movement

All animals are able to move parts of their bodies, even though some spend all their adult lives fixed in one place. The way animals move depends on their size and shape, and also on their surroundings. Land animals are able to push against solid ground, but animals that fly or swim push against moving air or water. To do this, most of these animals use wings or fins – but a few, such as octopuses and squid, use jet propulsion.

INSECT TAKE-OFF
A locust launches itself into the air using strong muscles in its legs, and with its wings flat against its body. Once in the air, the locust opens wide both of its pairs of wings and flaps them vigorously.

Hind legs push out strongly

Surface area of wings increased by feathers

During upstroke, broad wings sweep upwards

Bird has strong pectoral muscles to pull wings up

MOVING THROUGH WATER
A dogfish, like a shark, is a cartilaginous fish. It swims by curving its body into an S-shape. This motion propels the fish through the water, as its body and fins push the water aside. A bony fish swims in a different way. It keeps its body straight and beats its tail to move forwards.

Movement begins at the head and follows through body

Tail pushes against water at the end of each wave

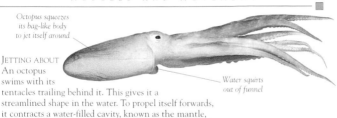

Octopus squeezes
its bag-like body
to jet itself around

Water squirts
out of funnel

JETTING ABOUT
An octopus
swims with its
tentacles trailing behind it. This gives it a
streamlined shape in the water. To propel itself forwards,
it contracts a water-filled cavity, known as the mantle,
and squirts a stream of water out of its body via a funnel.

A BIRD IN FLIGHT

MOVING THROUGH AIR
Birds, insects, and bats are the only animals capable of powered flight. They
use energy to flap their wings and push against gravity, so that they
stay airborne. A bird's wings are specially curved in a shape called an
aerofoil that helps to produce an upward lift. If the bird stops
flapping, it slows, lift decreases, and the bird starts to drop.

BLUE PIGEON

Muscles now
push wings
downwards

Downstroke pushes
bird up and forwards

Wings move
up for a new
upstroke

Wing
movement
produces lift to
counteract gravity

Dogfish body
bends in S-shape

Flat fins
maintain fish
at same level
in water

Head swings
around and
new curve starts

NERVOUS SYSTEM AND BRAIN

NERVE CELLS, OR NEURONS, carry signals from one part of an animal's body to another, so that it can work in a coordinated way. Together, neurons form a network called a nervous system. In many animals, this is controlled by the brain, which receives and processes information from the nerve cells.

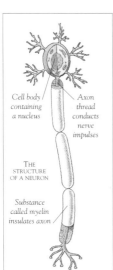

Cell body containing a nucleus

Axon thread conducts nerve impulses

THE STRUCTURE OF A NEURON

Substance called myelin insulates axon

CARRYING SIGNALS
A typical nerve cell has a cell body, with a nucleus, and an axon that carries signals. In vertebrates, the insulated axon helps signals travel quickly.

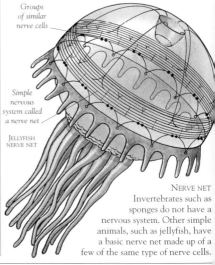

Groups of similar nerve cells

Simple nervous system called a nerve net

JELLYFISH NERVE NET

NERVE NET
Invertebrates such as sponges do not have a nervous system. Other simple animals, such as jellyfish, have a basic nerve net made up of a few of the same type of nerve cells.

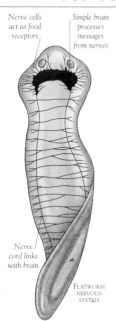

Nerve cells act as food receptors

Simple brain processes messages from nerves

Nerve cord links with brain

FLATWORM NERVOUS SYSTEM

BRAIN STRUCTURE

Cerebellum controls involuntary movements

Cerebrum

FROG BRAIN

The brains of vertebrates are divided into three main regions. The medulla controls vital processes, such as breathing. The cerebellum controls posture. The cerebrum deals with voluntary movements. A frog's cerebrum is quite small.

Medulla

BIRD BRAIN

A bird's brain has a large cerebellum that enables it to control the complex movements involved in flight. Its cerebrum is larger than a frog's but it is not folded.

Cerebrum   *Cerebellum*   *Medulla*

MAGNETIC RESONANCE IMAGE OF HUMAN BRAIN

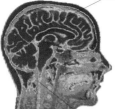

Folded cerebrum

HUMAN BRAIN

By far the biggest part of the human brain is the cerebrum. It is a large folded area of nerve cells split into right and left cerebral hemispheres. It governs memory, learning, and senses.

Medulla

BRAIN AND NERVE NET

A flatworm has a simple brain, which is connected to two nerve cords that run the length of its body. The nerve cords carry sensory signals to the brain. Once the brain has processed the signal, it makes the body respond.

SENSES

ANIMALS USE THEIR SENSES to gather information about the world around them. Many animals rely on their eyesight, but others depend on different senses. Some insects and mammals have highly developed hearing, and use sounds to build up an "image" of their surroundings.

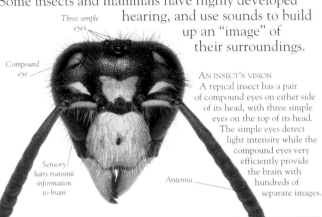

Three simple eyes

Compound eye

Sensory hairs transmit information to brain

Antenna

AN INSECT'S VISION
A typical insect has a pair of compound eyes on either side of its head, with three simple eyes on the top of its head. The simple eyes detect light intensity while the compound eyes very efficiently provide the brain with hundreds of separate images.

INSIDE A COMPOUND EYE
A compound eye has hundreds of compartments, each with a surface lens and an inner, conical lens. Each facet of the eye detects light from a very narrow area, and the brain puts all these images together.

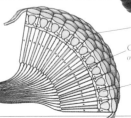

Facet

Compartment, or ommatidium

Cone-shaped lens inside compartment

CHAMELEON

Eye is able to swivel independently

DOUBLE VISION

When a chameleon spots its prey, it can keep one eye on it and still check all around for predators. This is because it is able to move its eyes independently. The chameleon's brain receives two separate images and must make sense of them. As it nears its prey, the chameleon focuses both eyes on it and takes careful aim.

WHITEFIN DOLPHIN

Chameleon has wide field of vision

Cricket's ears are near the knee joint on its front legs

ECHO-LOCATION

Some animals send out high-frequency sounds to put together a picture of the world. Dolphins and bats emit squeaks that bounce off surrounding objects – including prey – and return to the animal as echoes. They are interpreted by the animal's brain.

SENSING SOUNDS

Insects such as crickets and grasshoppers learn about the world mostly through their ability to detect vibrations. Crickets have "ears" on their knees made from a taut membrane that is sensitive to sound vibrations.

CENTRAL AMERICAN CRICKET

HEARING RANGE OF SELECTED ANIMALS	
Sound is measured by its pitch in units called Hertz (Hz). A higher Hertz number means a higher pitch – a lower number, a lower pitch.	
SPECIES	HEARING RANGES IN HZ
Elephant	1 – 20,000 Hz
Dog	10 – 35,000 Hz
Human	20 – 20,000 Hz
Bat	100 – 100,000 Hz
Frog	100 – 2,500 Hz

More senses

In addition to using their eyes
and ears, animals have special organs
to give them extra information about
their environment. Taste and smell are
similar senses that detect chemicals
and help animals find food. In the
water and in the dark, animals need
special senses to survive.

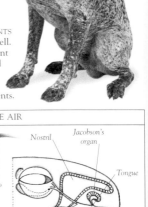

*Gundogs have
an excellent
sense of smell*

BRAQUE DU
BOURBONNAIS

DOG SCENTS
A dog's most advanced sense is its sense of smell.
Moisture on a dog's nose helps to dissolve scent
particles. Inside its muzzle is a large folded
area that traps scents and passes sensory
information to the brain. A large part of a
dog's brain is concerned with interpreting scents.

TASTING THE AIR

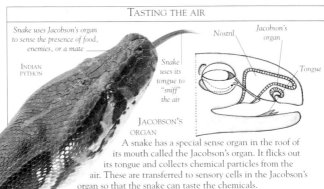

*Snake uses Jacobson's organ
to sense the presence of food,
enemies, or a mate*

INDIAN
PYTHON

*Snake
uses its
tongue to
"sniff"
the air*

Nostril

*Jacobson's
organ*

Tongue

JACOBSON'S
ORGAN
A snake has a special sense organ in the roof of
its mouth called the Jacobson's organ. It flicks out
its tongue and collects chemical particles from the
air. These are transferred to sensory cells in the Jacobson's
organ so that the snake can taste the chemicals.

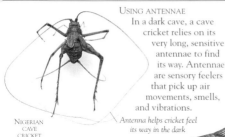

USING ANTENNAE

In a dark cave, a cave cricket relies on its very long, sensitive antennae to find its way. Antennae are sensory feelers that pick up air movements, smells, and vibrations.

NIGERIAN CAVE CRICKET

Antenna helps cricket feel its way in the dark

SENSE FACTS

• A dog's nose has 200 million scent receptors. A human nose has five million.

• A dragonfly can spot an insect moving 10 m (33 ft) away.

• Some bats can detect a tiny midge 20 m (65.5 ft) away.

LATERAL LINE

Most fish have a line of receptors on the sides of their bodies called a lateral line. This system detects changes in water pressure and signals to the fish that there is movement nearby.

MIRROR CARP

Large, shiny scales mark the lateral line

SENSING GRAVITY

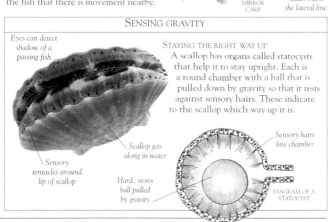

Eyes can detect shadow of a passing fish

STAYING THE RIGHT WAY UP

A scallop has organs called statocysts that help it to stay upright. Each is a round chamber with a ball that is pulled down by gravity so that it rests against sensory hairs. These indicate to the scallop which way up it is.

Sensory tentacles around lip of scallop

Scallop jets along in water

Hard, stony ball pulled by gravity

Sensory hairs line chamber

DIAGRAM OF A STATOCYST

BEHAVIOUR

THE THINGS THAT AN ANIMAL DOES and the way that it does them make up its behaviour. This may include the way an animal finds food, looks after itself, protects its territory, finds a partner, or cares for its young.

A penguin must keep its feathers waterproof

CARING FOR FEATHERS

To keep its feathers in good condition, a bird must spend time each day caring for them. It pushes its beak through ruffled feathers to smooth them out and remove parasites such as feather lice. Penguins and other water birds spread oil from a special gland through their feathers to keep them waterproof.

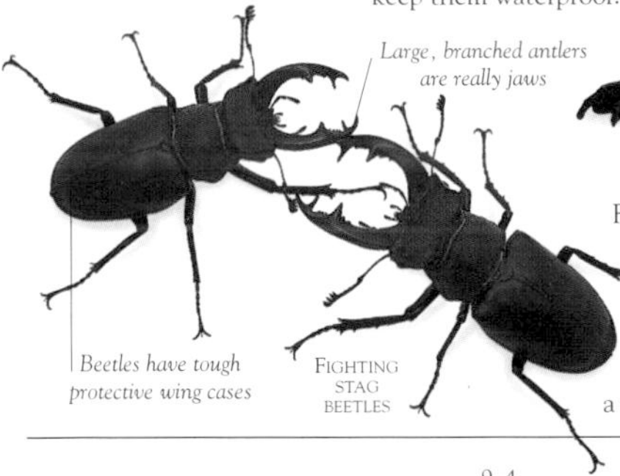

Special gland near the tail contains oil

KING PENGUIN

Large, branched antlers are really jaws

FIGHTING FOR TERRITORY

Some animals fight to defend their territory from competitors. Male stag beetles have powerful jaws that they use to grasp a rival and lift it out of the way.

Beetles have tough protective wing cases

FIGHTING STAG BEETLES

USING TOOLS

Some animals have learned to use a tool to help them find food. Sea otters, for example, anchor themselves in a seaweed called kelp, and use a stone from the sea bed to break open shellfish. They rest the stone on their chest and smash the shellfish against it. They can then eat the contents.

BIRD MIGRATION

ARCTIC TERN

Some birds fly long distances (migrate) to avoid bad weather or find food. Arctic terns fly from the Arctic to the Antarctic.

AMERICAN GOLDEN-PLOVER

This bird makes the longest migration of any land bird. It breeds in Canada and flies south to Argentina for the winter.

SHORT-TAILED SHEARWATER

After breeding in the south of Australia, this shearwater flies around the North Pacific and back again.

GREATER WHITETHROAT

The greater whitethroat is a small warbler that breeds during spring and summer in Europe. It migrates to Africa for winter.

Communication and defence

To survive, animals have to communicate with their own kind, and defend themselves against predators. Animals communicate in many ways. Some use sounds, while others use visual signals or chemical messages. Animal defences include camouflage, armour, and poisons.

INDIAN MOON MOTH

CHEMICAL COMMUNICATION
Many insects communicate using chemicals called pheromones that they release onto the ground or into the air. The female Indian moon moth makes a pheromone that attracts males from far away.

Moths are able to respond to scents

Bee performs waggle dance in the hive

HONEYBEE DANCE
When a honeybee locates a good source of nectar, it will return to the hive and inform other bees of its find. The bee performs a waggling dance that surrounding bees are able to interpret. It tells them the location, distance, and quality of the food.

ROBIN IN SONG

BIRDSONG
Animals such as birds and frogs communicate with sounds. A bird has a chamber called a syrinx that can produce a range of sounds to warn other birds to stay out of its territory.

CAMOUFLAGED INSECT

This spiny stick insect relies on camouflage to escape attack. It hides among foliage, where its body resembles a cluster of dead leaves. This disguise makes it very difficult to find, but works only if the insect remains still.

GIANT SPINY STICK INSECT

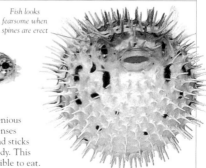

Brightly coloured frog is easily seen in the forest

POISONOUS YELLOW MANTELLA

POISON FROG

Some frogs contain poisons in their skin that protect them from predators. They have brightly coloured skin to warn predators that they are poisonous.

While stick insect stays still, it is hidden in the leaves

PORCUPINE FISH

When the fish relaxes, its spines lie flat against its body

Fish looks fearsome when spines are erect

PUFFED UP

The porcupine fish has an ingenious method of defence. When it senses danger, it fills up with water and sticks out the spines that cover its body. This makes the fish virtually impossible to eat.

REPRODUCTION

ALL LIVING THINGS REPRODUCE. Animals do this in two ways. For most species, two parents come together, in a process called mating, to create a fertilized egg. In a few very simple animals, one individual reproduces itself.

Peacocks fan out their tail feathers in a display to the peahen

PEACOCK

COURTSHIP
Before an animal can reproduce, it must first find a partner of the opposite sex. To attract a mate, male animals put on a display and act out a ritual known as courtship. This complex behaviour forms a bond between the pair.

MALE FROG FERTILIZES EGGS

The male frog sprays the eggs with his sperm

ROD-SHAPED BACTERIA ESCHERICHIA COLI

Offspring are identical to the parents

ASEXUAL REPRODUCTION
Many simple forms of life, such as bacteria, reproduce by splitting in two. Animals such as aphids produce new individuals from an unfertilized egg.

SEXUAL REPRODUCTION
Most reproduction requires that a male and a female parent come together so that the female's sex cells are fertilized. A frog's eggs are fertilized externally but in some animals the male's sperm enters the female to reach the egg cells inside her body.

REPRODUCTION RATES

Some animals can reproduce extremely rapidly, but only a few of their offspring survive to become adults.

SPECIES	BREEDING AGE	OFFSPRING PER YEAR
Northern gannet	5–6 years	1
Rabbit	8 months	10–30
Nile crocodile	15 years	50
Fruit fly	10–14 days	Up to 900

DOGFISH
EGG CASES

Egg case
hung by
tendrils
from
seaweed

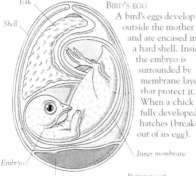

Yolk

Shell

Embryo

Air sac

Inner membrane

BIRD'S EGG
A bird's eggs develop outside the mother and are encased in a hard shell. Inside, the embryo is surrounded by membrane layers that protect it. When a chick is fully developed it hatches (breaks out of its egg).

EGG CASES
The female dogfish releases rubbery egg cases, each containing an embryo. The embryo grows inside the case, living on yolk that acts as a food store. The fish emerges after 6 to 9 months.

BURMESE CAT
WITH KITTENS

*Kittens born at
the same time may
look different*

CARING FOR YOUNG
Mammals such as cats give birth to live young that are helpless for the first few weeks. The young rely on their mother's milk until they are strong enough to find their own food.

GROWTH AND DEVELOPMENT

AS LIVING THINGS GET OLDER they usually grow bigger.
Parts of their bodies grow at different rates, so that
their proportions gradually alter. Some animals grow
all their lives; others develop rapidly when they are
young, then stop when they are mature. Some animals
can grow a new body part to replace a lost one.

GROWING A NEW TAIL

TREE SKINK
If a predator catches a tree skink by its tail, the tail breaks off at a fracture point,
letting the skink escape. A new tail gradually grows, containing rubbery cartilage
rather than bone. Eight months later, the tail has regrown to
its original length.

Tail shed at fracture point

Growing new tail uses a lot of energy

1 STARTING LIFE
A hen's egg starts life as a single cell.
Once it has been fertilized, the cell
divides many
times and
these new
cells form
the chick's
body.

Chick uses egg tooth to crack shell

2 BREAKING FREE
During hatching, the chick pecks
through the shell, and kicks its way out.
It now breathes
fresh air for the
first time.

Wet down feathers soon dry

GROWING A SHELL

Many molluscs grow hard shells by depositing crystals of calcium carbonate. As the mollusc grows, its shell grows too. Bands in the shell show each season's growth.

TRITON SHELL
GROWTH

Whirls become larger and more pronounced

A fully developed shell has a thick lip

GROWING UP

Newborn cub is vulnerable

Eight-week-old cub is more independent

At 12 weeks, cub play-fights, preparing to catch its own prey

FOX CUB

A newborn fox cub is helpless for the first few weeks of life and its mother must take care of it. At eight weeks, it can walk and find its own food, though the mother still provides milk.

Chicks are able to find their own food

4 FOLLOWING THEIR MOTHER
The chicks now have dry down and can walk. They recognize their mother's call and soon recognize her by sight as well.

3 NEWBORN
The young chick has a well-developed head and feet but its wings are immature. It can run around and will peck at the ground looking for food.

METAMORPHOSIS

As animals grow, their bodies change shape. Some animals change slightly, but the young of amphibians, fish, and insects look very different from the adults. This process of change is called metamorphosis. Insects change shape in two ways – either by complete metamorphosis (changing shape suddenly), or by incomplete metamorphosis (changing by gradual stages).

COMPLETE METAMORPHOSIS: THE LADYBIRD

Eggs laid on the underside of leaves

Soft, pale larva emerges

Adult larva feeds on aphids

1 THE EGG
A ladybird changes shape completely as it grows up. The egg takes about four days to hatch.

2 LARVA EMERGES
A soft-bodied larva hatches from the egg. Its first meal is its egg, which contains nutrients.

3 MATURE LARVA
The mature larva walks around slowly and cannot fly. Its colours warn off predators.

Larval skin is shed and pupal skin forms

Wing cases gradually change colour

Adult ladybird feeds on aphids

4 FORMING A PUPA
After eating enough aphids, the larva attaches itself to a leaf, ready to pupate.

5 YOUNG ADULT
A soft, yellow-winged ladybird hatches out of the pupa. Its wing cases start to harden.

6 MATURE LADYBIRD
The adult ladybird looks very different from its larva. It has its adult colours and is able to fly.

INCOMPLETE METAMORPHOSIS: THE DAMSELFLY

1 NYMPH
A damselfly nymph lives underwater, but carries out its final moult out of the water.

2 PULLING FREE
Blood pumped into the thorax makes it expand and burst out of its nymphal skin.

3 OLD SKIN
The adult head pulls out of the nymphal skin and leaves its mask behind.

4 SOFT WINGS
Free of its old skin, blood fills the damselfly's wings, making them longer.

5 GROWING
Thorax and abdomen are still growing. The wings are delicate.

6 ABLE TO FLY
The adult is ready to fly. Its abdomen is long and shiny and its wings are transparent.

INSECT METAMORPHOSIS

Insects have three or four stages in their life-cycles, depending on whether they undergo complete or incomplete metamorphosis. The stages are often very different in length. The first section of this chart shows the length of time needed for the four stages of complete metamorphosis.

Species	Egg	Larva	Pupa	Adult
Bluebottle	1 day	8 days	9 days	35 days
Ladybird	4 days	18 days	15 days	9 months
Large white butterfly	14 days	1 month	6 months	2 months

INCOMPLETE METAMORPHOSIS IN INSECTS

Species	Egg	Nymph		Adult
Periodical cicada	1 month	17 years	–	2 months
Mayfly	1 month	3 years	–	1 day
Cockroach	1 month	3 months	–	9 months

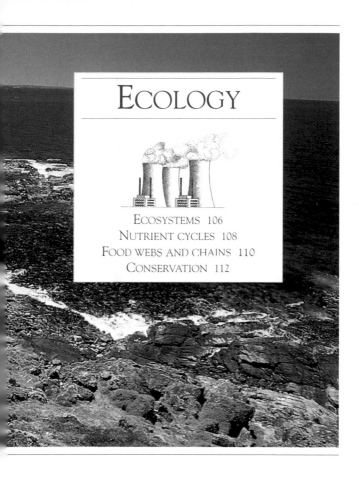

ECOLOGY

ECOSYSTEMS

LIVING THINGS DO NOT EXIST IN ISOLATION; they constantly interact with each other and with their environment. The study of these interactions is called ecology. Groups of organisms and their surroundings make up separate ecosystems. An ecosystem can be a pond, a forest, a beach, or an entire mountainside.

UNIQUE PLANET
Earth is the only planet known to support life. Its atmosphere contains the elements that are essential to sustain life and to protect us from the harmful effects of the Sun's rays.

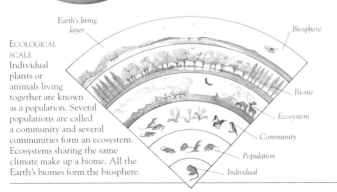

ECOLOGICAL SCALE
Individual plants or animals living together are known as a population. Several populations are called a community and several communities form an ecosystem. Ecosystems sharing the same climate make up a biome. All the Earth's biomes form the biosphere.

Earth's living layer

Biosphere

Biome

Ecosystem

Community

Population

Individual

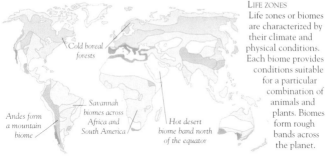

Cold boreal forests

Savannah biomes across Africa and South America

Andes form a mountain biome

Hot desert biome band north of the equator

LIFE ZONES

Life zones or biomes are characterized by their climate and physical conditions. Each biome provides conditions suitable for a particular combination of animals and plants. Biomes form rough bands across the planet.

- ○ TUNDRA
- ○ TEMPERATE FOREST
- ○ SAVANNAH
- ○ TEMPERATE GRASSLAND
- ○ TEMPERATE RAINFOREST
- ○ BOREAL FOREST
- ○ DESERT
- ○ TROPICAL RAINFOREST
- ○ MOUNTAIN
- ○ SCRUBLAND

EXAMPLES OF BIOMES

TUNDRA
Around the Arctic region are areas of tundra. Here the soil stays frozen for most of the year and only mosses and lichens grow. The climate is very dry.

TEMPERATE FOREST
Northern Europe and North America are known as temperate – their climates are neither very hot nor very cold. Deciduous trees flourish in these areas.

NUTRIENT CYCLES

ALL LIVING THINGS need chemical nutrients. These include carbon, oxygen, nitrogen, and water, and trace elements such as copper and zinc. Chemical nutrients are constantly recycled through the Earth's biosphere – they pass between living and non-living things.

Carbon dioxide is given out by green plants during respiration

Animals breathe out carbon dioxide

Animals take in carbon from plants

Carbon dioxide absorbed by plants during photosynthesis

Bacteria break down or decompose dead matter and produce carbon dioxide

Dead plants and animals decay

NUTRIENT CYCLE FACTS

• At least 25 of the Earth's 90 elements are used by living things.

• Plants absorb one-tenth of the atmosphere's carbon every year.

• Some parts of a nutrient cycle can take seconds, while others take thousands of years.

CARBON CYCLE

Carbon is a part of all living things. It moves around the living world in a constant cycle. Plants absorb carbon dioxide from the atmosphere during photosynthesis and animals take in carbon when they eat plants. Carbon is released when plants and animals decompose.

OXYGEN CYCLE
During photosynthesis, all plants release oxygen. Living things take in oxygen to break down the energy in their food.

NITROGEN CYCLE
Nitrogen is needed by living things to make proteins but it must be combined with other elements before it can be used. Some nitrogen is combined by lightning, but most is combined by bacteria that live in soil.

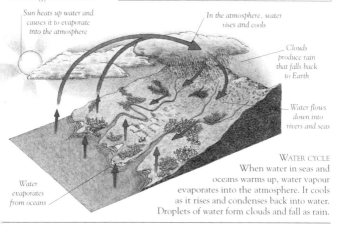

Sun heats up water and causes it to evaporate into the atmosphere

In the atmosphere, water rises and cools

Clouds produce rain that falls back to Earth

Water flows down into rivers and seas

Water evaporates from oceans

WATER CYCLE
When water in seas and oceans warms up, water vapour evaporates into the atmosphere. It cools as it rises and condenses back into water. Droplets of water form clouds and fall as rain.

FOOD WEBS AND CHAINS

IN ANY COMMUNITY, living things are linked together in food chains or webs. Energy is passed along the chain in the form of food. At the base of the chain are the primary producers, usually plants, which make their own food. When animals, or consumers, eat plants, energy is passed on, though a great deal is lost along the way. Poisons can also be passed through the chain.

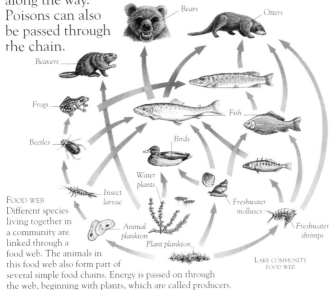

Bears

Otters

Beavers

Frogs

Beetles

Fish

Birds

Water plants

Insect larvae

Freshwater molluscs

Animal plankton

Plant plankton

Freshwater shrimps

LAKE COMMUNITY FOOD WEB

FOOD WEB
Different species living together in a community are linked through a food web. The animals in this food web also form part of several simple food chains. Energy is passed on through the web, beginning with plants, which are called producers.

Rabbit is primary consumer

Plant is primary producer

WOODLAND FOOD CHAIN

FOOD CHAIN

A food chain contains living things that provide food for each other. The chain usually has only three or four links because energy is lost at each stage. The first stage of the chain is typically a plant.

Fox is secondary consumer

TROPHIC LEVELS

Ecologists call each stage of a food chain a trophic level. There is more living material and more energy available at the first trophic level, the primary producers. Energy is used up by organisms at each successive layer of consumption so that less can be stored and passed on.

TROPHIC LEVELS PYRAMID

At this trophic level, there are fewer consumers

Less energy available at this level

Energy is used up by living things

Primary producers

BELUGAS' POISON CHAIN
Belugas in North America are part of a river food chain. Toxins pumped into the river are absorbed by plankton, then passed to fish that eat plankton, and then to whales that eat fish. The toxins have become concentrated in the whales.

CONSERVATION

HUMAN ACTIVITY has a great impact on the environment. Humans use up the Earth's resources and leave behind a great deal of waste and pollution. To protect the environment, we are learning to practise conservation and to manage our use of natural resources.

HABITAT CHANGE
Humans have drastically altered the natural world. Large amounts of land have been cleared to make way for the increased demand for houses and roads. As a result, the balance of the ecosystems in those areas is upset.

CLIMATE CHANGE
A rise of a few degrees in the Earth's temperature has a significant effect on the environment. It can cause rivers, such as this one in Spain, to dry up, threatening freshwater fish and amphibians, and other animals that depend on the river.

Trawlers locate and process huge amounts of fish

OVER-FISHING
Technological advances in fishing have led to greater yields. Huge trawlers now catch such vast amounts of fish that the fish population in some areas has been severely reduced. Fishing bans have been enforced in an effort to help fish stocks recover.

REINTRODUCTION

In the 1970s, the red kite was on the verge of extinction and a programme was set up to reintroduce the bird to Britain. In programmes such as this one, young animals are raised in captivity and then set free. They are released in special areas where their progress can be carefully monitored.

CONSERVATION FACTS

• Since 1700, 200 species or subspecies of birds have become extinct.

• Over 1,000 of the world's bird species and 500 of its mammals are currently endangered.

• Conservation has saved species such as the grey whale and Hawaiian goose.

ALLIGATORS

Wild animals sometimes need to be protected by law. Alligators used to be hunted for their skins and became endangered. The alligators of the southern US are now protected by a law that limits hunting.

NATURE RESERVES

All over the world, nature reserves have been set up. This one in Kenya protects animals, such as rhinos, from poachers. The reserves preserve the environment and the animals, and attract tourists whose money helps maintain the reserves.

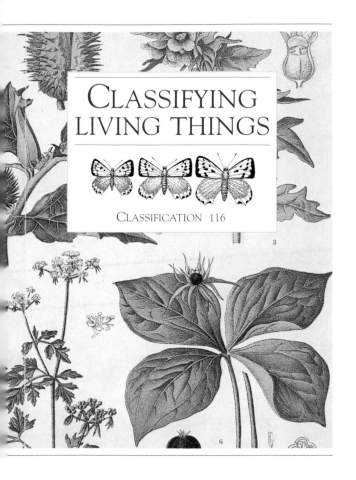

CLASSIFYING LIVING THINGS

CLASSIFICATION 116

3

6

CLASSIFICATION 1

BIOLOGISTS USE CLASSIFICATION to identify types of living things, and show how they are related through evolution. Classification details and species totals alter as more is discovered about the living world. This scheme is based on five kingdoms. It includes the most important categories in each group.

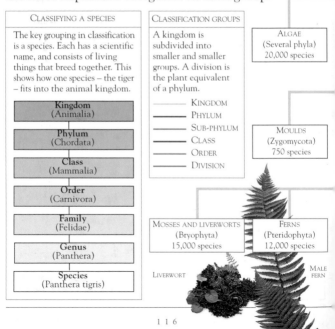

CLASSIFYING A SPECIES

The key grouping in classification is a species. Each has a scientific name, and consists of living things that breed together. This shows how one species – the tiger – fits into the animal kingdom.

| Kingdom (Animalia) |
| Phylum (Chordata) |
| Class (Mammalia) |
| Order (Carnivora) |
| Family (Felidae) |
| Genus (Panthera) |
| Species (Panthera tigris) |

CLASSIFICATION GROUPS

A kingdom is subdivided into smaller and smaller groups. A division is the plant equivalent of a phylum.

- KINGDOM
- PHYLUM
- SUB-PHYLUM
- CLASS
- ORDER
- DIVISION

ALGAE (Several phyla) 20,000 species

MOULDS (Zygomycota) 750 species

MOSSES AND LIVERWORTS (Bryophyta) 15,000 species

FERNS (Pteridophyta) 12,000 species

LIVERWORT

MALE FERN

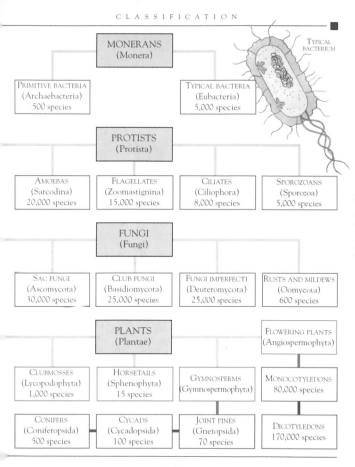

MONERANS
(Monera)

TYPICAL BACTERIUM

PRIMITIVE BACTERIA
(Archaebacteria)
500 species

TYPICAL BACTERIA
(Eubacteria)
5,000 species

PROTISTS
(Protista)

AMOEBAS
(Sarcodina)
20,000 species

FLAGELLATES
(Zoomastignina)
15,000 species

CILIATES
(Ciliophora)
8,000 species

SPOROZOANS
(Sporozoa)
5,000 species

FUNGI
(Fungi)

SAC FUNGI
(Ascomycota)
30,000 species

CLUB FUNGI
(Basidiomycota)
25,000 species

FUNGI IMPERFECTI
(Deuteromycota)
25,000 species

RUSTS AND MILDEWS
(Oomycota)
600 species

PLANTS
(Plantae)

FLOWERING PLANTS
(Angiospermophyta)

CLUBMOSSES
(Lycopodophyta)
1,000 species

HORSETAILS
(Sphenophyta)
15 species

GYMNOSPERMS
(Gymnospermophyta)

MONOCOTYLEDONS
80,000 species

CONIFERS
(Coniferopsida)
500 species

CYCADS
(Cycadopsida)
100 species

JOINT PINES
(Gnetopsida)
70 species

DICOTYLEDONS
170,000 species

Classification 2

Animals have been studied more closely than any other forms of life. Biologists have identified and classified nearly all the species of vertebrates (animals with backbones) although it is likely that new species of fish await discovery. By contrast, the world of invertebrates is not so well documented and there may be many more species to be identified.

Springtails
Bristletails
Diplurans
Silverfish
Mayflies
Stoneflies
Webspinners
Dragonflies
Grasshoppers, crickets
Stick and leaf insects
Grylloblattids
Earwigs
Cockroaches
Praying mantids
Termites
Lice
Thrips
Booklice
Zorapterans
Bugs
Beetles
Ants, bees, wasps
Lacewings and antlions
Scorpionflies
Stylopids
Caddisflies
Butterflies and moths
Flies
Fleas

INSECTS
(Insecta)
1,000,000 species

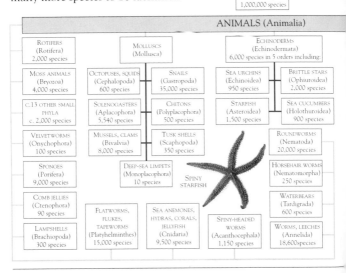

ANIMALS (Animalia)

ROTIFERS
(Rotifera)
2,000 species

MOSS ANIMALS
(Bryozoa)
4,000 species

C.13 OTHER SMALL PHYLA
c. 2,000 species

VELVETWORMS
(Onychophora)
100 species

SPONGES
(Porifera)
9,000 species

COMB JELLIES
(Ctenophora)
90 species

LAMPSHELLS
(Brachiopoda)
300 species

MOLLUSCS
(Mollusca)

OCTOPUSES, SQUIDS
(Cephalopoda)
600 species

SOLENOGASTERS
(Aplacophora)
5,540 species

MUSSELS, CLAMS
(Bivalvia)
8,000 species

DEEP-SEA LIMPETS
(Monoplacophora)
10 species

SNAILS
(Gastropoda)
35,000 species

CHITONS
(Polyplacophora)
500 species

TUSK SHELLS
(Scaphopoda)
350 species

ECHINODERMS
(Echinodermata)
6,000 species in 5 orders including:

SEA URCHINS
(Echinoidea)
950 species

STARFISH
(Asteroidea)
1,500 species

BRITTLE STARS
(Ophiuroidea)
2,000 species

SEA CUCUMBERS
(Holothuroidea)
900 species

ROUNDWORMS
(Nematoda)
20,000 species

HORSEHAIR WORMS
(Nematomorpha)
250 species

WATERBEARS
(Tardigrada)
600 species

WORMS, LEECHES
(Annelida)
18,600 species

SPINY STARFISH

FLATWORMS, FLUKES, TAPEWORMS
(Platyhelminthes)
15,000 species

SEA ANEMONES, HYDRAS, CORALS, JELLYFISH
(Cnidaria)
9,500 species

SPINY-HEADED WORMS
(Acanthocephala)
1,150 species

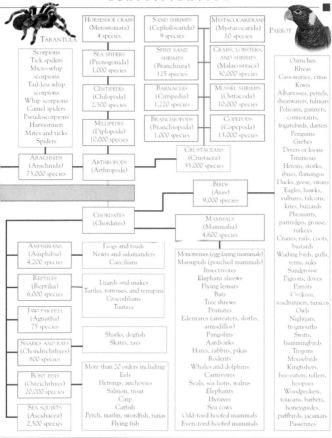

TARANTULA

PARROT

	HORSESHOE CRABS (Merostomata) 4 species	SAND SHRIMPS (Cephalocarida) 9 species	MYSTACOCARIDEANS (Mystacocarida) 10 species

Scorpions
Tick spiders
Micro-whip scorpions
Tail-less whip scorpions
Whip scorpions
Camel spiders
Pseudoscorpions
Harvestmen
Mites and ticks
Spiders

SEA SPIDERS (Pycnogonida) 1,000 species

CENTIPEDES (Chilopoda) 2,500 species

MILLIPEDES (Diplopoda) 10,000 species

ARACHNIDS (Arachnida) 73,000 species

ARTHROPODS (Arthropoda)

SPINY SAND SHRIMPS (Branchiura) 125 species

BARNACLES (Cirripedia) 1,220 species

BRANCHIOPODS (Branchiopoda) 1,000 species

CRABS, LOBSTERS, AND SHRIMPS (Malacostraca) 30,000 species

MUSSEL SHRIMPS (Ostracoda) 10,000 species

COPEPODS (Copepoda) 13,000 species

CRUSTACEANS (Crustacea) 35,000 species

CHORDATES (Chordates)

BIRDS (Aves) 9,000 species

MAMMALS (Mammalia) 4,600 species

AMPHIBIANS (Amphibia) 4,200 species

Frogs and toads
Newts and salamanders
Caecilians

REPTILES (Reptilia) 6,000 species

Lizards and snakes
Turtles, tortoises, and terrapins
Crocodilians
Tuatara

JAWLESS FISH (Agnatha) 75 species

SHARKS AND RAYS (Chondrichthyes) 800 species

Sharks, dogfish
Skates, rays

BONY FISH (Osteichthyes) 20,000 species

SEA SQUIRTS (Ascidiacea) 2,500 species

More than 20 orders including:
Eels
Herrings, anchovies
Salmon, trout
Carp
Catfish
Perch, marlin, swordfish, tunas
Flying fish

Monotremes (egg-laying mammals)
Marsupials (pouched mammals)
Insectivores
Elephant shrews
Flying lemurs
Bats
Tree shrews
Primates
Edentates (anteaters, sloths, armadillos)
Pangolins
Aardvarks
Hares, rabbits, pikas
Rodents
Whales and dolphins
Carnivores
Seals, sea lions, walrus
Elephants
Hyraxes
Sea cows
Odd-toed hoofed mammals
Even-toed hoofed mammals

Ostriches
Rheas
Cassowaries, emus
Kiwis
Albatrosses, petrels, shearwaters, fulmars
Pelicans, gannets, cormorants, frigatebirds, darters
Penguins
Grebes
Divers or loons
Tinamous
Herons, storks, ibises, flamingos
Ducks, geese, swans
Eagles, hawks, vultures, falcons, kites, buzzards
Pheasants, partridges, grouse, turkeys
Cranes, rails, coots, bustards
Wading birds, gulls, terns, auks
Sandgrouse
Pigeons, doves
Parrots
Cuckoos, roadrunners, turacos
Owls
Nightjars, trogmouths
Swifts, hummingbirds
Trogons
Mousebirds
Kingfishers, bee-eaters, rollers, hoopoes
Woodpeckers, toucans, barbets, honeyguides, puffbirds, jacamars
Passerines

Glossary

ABDOMEN
An animal body part; in an insect, it contains the digestive system and reproductive organs.

ACHENE
A dry fruit that has a single seed.

ALGAE (singular ALGA)
Simple, plant-like organisms that create their own food by photosynthesis and live in groups or colonies, often in water.

AMOEBA
A tiny, single-celled, jelly-like protozoan.

AMPHIBIAN
Cold-blooded vertebrate that lives on land but usually breeds in water.

ANTENNA
A sensory feeler on the head of an arthropod.

ANTHER
A part of the male reproductive organs of a flower that produces pollen.

ARACHNID
An arthropod, such as a spider or scorpion, with four pairs of walking legs.

ARTHROPOD
An invertebrate, with a jointed body case, including crustaceans, arachnids, and insects.

AXON
A long, thread-like part of a nerve cell or neuron.

BACTERIA (singular BACTERIUM)
Single-celled, microscopic organisms.

BARK
The strong, outer skin of woody plants.

BIOME
An ecological zone characterized by climate and vegetation.

BIOSPHERE
The layer of the Earth, including its atmosphere, where living things are found.

BULB
A plant's underground food store.

CARNIVORE
An animal whose diet consists mainly of meat.

CARPEL
A flower's female reproductive organ, with ovary, stigma, and style.

CARTILAGE
A strong, flexible, and slippery substance in the skeleton of a vertebrate.

CELL
A tiny unit that makes up all living things.

CHLOROPLAST
A structure in a plant cell that contains chlorophyll.

CHRYSALIS
A hard, shiny case in which an insect changes shape during complete metamorphosis.

COMPOUND EYE
An eye made of many separate compartments, each with its own lens.

CONIFER
Plant that reproduces using cones; often grows in cool areas.

CONSERVATION
Managing the planet's resources to protect the natural world.

CRUSTACEAN
Typically, an aquatic arthropod with jointed legs and a hard, outer body case.

CYTOPLASM
Jelly-like substance
inside a cell.

DECIDUOUS
A tree or shrub that
sheds all its leaves at
the same time.

DEHISCENT FRUIT
A dry fruit that splits
open to release its seeds.

DNA
Deoxyribonucleic acid,
a substance in all
living things that
stores information.

ECHOLOCATION
Using high-frequency
sounds to make sense
of surroundings.

ECOLOGY
The study of the
relationships between
living things and their
environments.

ECOSYSTEM
A community of living
things and their
surroundings.

EVOLUTION
The process by which
organisms gradually
change from one
generation to the next.

EXOSKELETON
A tough skeleton
that surrounds an
animal's body.

FERTILIZATION
The process whereby
male and female sex
cells join together during
sexual reproduction.

FLOWER
Part of a flowering
plant that contains
the male and female
reproductive organs.

FOSSIL
The preserved remains
of an organism.

FRUIT
A fruit contains a
plant's seeds and may
be hard and dry or
soft and juicy.

FUNGI (singular FUNGUS)
An organism that
absorbs food from
its surroundings.

GERMINATION
When a seed or spore
begins to grow.

GESTATION
The development of
the embryo inside
a mammal.

GILL
A thin flap that allows
water-dwelling animals
to breathe.

GYMNOSPERM
A plant whose seeds
do not develop inside
a fruit.

HABITAT
The place where a
species is usually found.

HOMINID
A family of primates
including humans and
their direct ancestors.

HOMINOID
A group that includes
hominids and apes.

INFLORESCENCE
A group of flowers
on a stem.

INVERTEBRATE
An animal that does
not have a backbone.

JACOBSON'S ORGAN
A sensory organ in
reptiles that detects
chemicals in the air.

LARVA
An immature animal
that changes during the
process of complete
metamorphosis into
an adult form.

LEGUME
A dry fruit, such as a
pea pod, that splits
open along two sides.

LICHEN
A partnership between
an alga and a fungus.

MAMMAL
An animal with hair
on its body that feeds
its young on milk.

MARSUPIAL
A mammal whose young develop inside their mother's pouch.

MESOPHYLL
The cells inside a leaf.

METAMORPHOSIS
A change in body shape in some invertebrates and amphibians that can be either "complete" or "incomplete".

MIGRATION
The movement of many birds, mammals, fish, and insects to a more favourable habitat at certain times of the year.

MOLLUSC
An invertebrate with a soft body that is often covered by a hard shell.

MOULT
When an animal sheds its outer covering (skin, feathers, or fur) from time to time, usually in order to replace it.

MUSCLE
A bundle of cells that are able to contract and relax to produce movement.

MYCELIUM
A network of feeding threads called hyphae, produced by fungi.

NERVE
A bundle of cells that carries signals.

NERVOUS SYSTEM
A network of nerve cells.

NUCLEUS
The control centre of a cell.

NYMPH
A young insect that matures by incomplete metamorphosis.

ORGAN
A structure in a plant or animal that carries out particular functions.

ORGANELLE
A structure in a cell that performs a specific task.

ORGANIC
Related to living things.

ORGANISM
A living animal, plant, fungus, or bacterium.

OVARY
A female reproductive organ that produces ova (eggs) in an animal, or contains ovules (with female sex cells) in a plant.

OVULE
A structure in a plant that contains female sex cells. After fertilization, the ovule develops into the seed.

PARASITE
An organism that lives in or on another organism (a host).

PETAL
A flower's inner flap that may help to attract animal pollinators to plants.

PHEROMONE
A chemical given out by an animal that has an effect on another.

PHLOEM
A group of cells that carry nutrients around a plant.

PLACENTA
An organ in mammals that allows substances to pass between the mother and her unborn young.

POLLEN
Microscopic grains containing the male sex cells in plants.

POLLINATION
The process of transferring pollen from the anthers of one flower to the stigma of another.

POLYP
An invertebrate with a cylindrical body and tentacles that is attached to a solid structure.

PRIMATE
A mammal with flexible fingers and eyes that point forwards. Humans are primates.

PROSIMIAN
A primitive tree-dwelling primate, such as a lemur or loris.

PROTOZOA
A group of single-celled organisms that have to take in food, such as amoebas.

PUPA
The stage during complete metamorphosis when the insect changes its shape.

REPTILE
A cold-blooded, scaly-skinned vertebrate that lays leathery shelled eggs.

RHIZOID
A simple, root-like thread of mosses and liverworts.

ROOT
The part of a plant that holds it to the ground and supplies it with water and minerals.

SEED
A seed contains the beginning of a new plant and its food store.

SEPAL
A flower's outer flap.

SHOOT
The stem, leaves, and flowers of a plant.

SKELETON
A supporting framework of an animal's body.

SPECIES
A group of living things in the natural world that are able to breed together.

SPIRACLE
A hole in the body of an insect through which air enters.

STAMEN
A flower's male reproductive organ.

STATOCYST
An organ that detects the pull of gravity in invertebrates.

STIGMA
A part of the female reproductive organs of a flower that collects the pollen.

STOMATA (SINGULAR STOMA)
Tiny slits on leaf's surface that open and close.

STYLE
A stalk that connects the stigma to the ovary in a flower.

SWIM BLADDER
An organ filled with gas that helps a fish to float in water.

THORAX
An animal body part; in an insect, it contains the legs and wings.

THYLAKOID
A disc-shaped sac inside a chloroplast that contains chlorophyll.

TRANSPIRATION
The movement of water up through a plant.

TUBER
An underground food store for a plant.

VACUOLE
A space inside a cell used for storage.

VERTEBRATE
An animal with a skeleton, including mammals, fish, reptiles, birds, and amphibians.

VIRUS
A package of chemicals that breaks into another cell and relies on it in order to reproduce.

XYLEM
A group of cells that transports water and minerals from the roots to the leaves.

Index

Acknowledgements

Dorling Kindersley would like to thank:
Alison McKittrick for additional picture
research; Lynn Bresler for the index;
Caroline Potts for picture library services;
Steve Wong for design assistance.

Photographers:
Peter Anderson; Geoff Brightling;
Jane Burton; Peter Chadwick; Andy
Crawford; Geoff Dann; Philip Dowell;
Neil Fletcher; Frank Greenaway; Colin
Keates; Dave King; Cyril Laubscher;
Mike Linley; Tracy Morgan; Andrew
McRobb; Roger Phillips; Tim Ridley;
Karl Shone; Kim Taylor; Andreas Von
Eisliedel; Jerry Young.

Illustrators:
Richard Bonson; Peter Bull; Angelica
Elsebach; Simone End (Linden Artists);
Robert Garwood; Will Giles; Mike Grey;
Stanley Johnson; Kenneth Lilly; Janosh
Marfy; Andrew McDonald; Richard Orr;
Sandra Pond; R. Turvey; John
Woodcock; Dan Wright.

Picture credits: t = top b = bottom
c = centre l = left r = right
The publisher would like to thank the
following for their kind permission to
reproduce the photographs:

Bruce Coleman: Erwin and Peggy Bauer
85bc; Nigel Blake 29tl; John Cancalosi
109tr, 112tl, 67 tr, Alain Compost 43tl,
63br, Gerald Cubitt 69 tr, Geoff Dore
107br, Peter Evans 96br, MPL Fogden
33tr, Jeff Foot 111cr, 95tr, David Hughs
85tr, Stephen J Krasemann 107bl, Luiz
Claudio Marigo 48/49, Hans Reinhard
113tl, 24/25, John Visser 63tr, Jean-
Pierre Zwaenepoel 72tl; Ecoscene: Pat
Groves 112cr; Mary Evans Picture
Library: Robert Francis 114/115; Garden
Picture Library: Brian Carter 35tl;
Robert Harding Picture Library: Robert
Francis 104/105; NASA: 106cl; Natural
History Museum: Peter Chadwick 37tr,
Philip Dowell 65tr, 75tr, Colin Keates
10/11, 33cr, 55cl, Dave King 53 cr, Karl
Shone 100cl, 100cr; Scarlett O'Hara 69
cl, 109tl; Oxford Scientific Films: 37br;
Planet Earth Pictures: JB Duncan 112bl;
Robert A. Jureit 33tc; Mike Potts 113bl;
Peter Scoones 17tl; Science Photo
Library: Dr Jeremy Burgess 38bc, 45tl,
P. Hawtin, University of Southampton
26bl, Manfred Kage 28bl, Mehau Kelyk
89br, John Reader 21cl; Tony Stone
Images: Peter Lambreti 113br;
Weymouth Sea Life Centre: Frank
Greenaway 87tr; Jerry Young: 59tr, 61cl,
67c, 84c.

Every effort has been made to trace the
copyright holders. Dorling Kindersley apologises
for any unintentional omissions and would be
pleased, in such circumstances, to add an
acknowledgement to future editions.